And The Sea Shall Have Them All

And The Sea Shall Have Them All

—

Art Milmore

Copyright 2004 Art Milmore
All rights reserved.

ISBN: **1539699080**
ISBN 13: **9781539699088**

Dedication

I Dedicate this book to the memory or my three mentors from the World War 2 generation in the order in which I knew them.

Mr. Leo A. DeLorey
- Carpenter, cabinet maker, boat builder, and metal fabricator

Mr. Edward Rowe Snow
- Author, athlete, treasure hunter, explorer, historian, teacher, and preservationist.

Mr. Gunter Blatt
-Tailor, photographer, holocaust survivor

Preface and Author Bio

—

FIFTY-SEVEN YEARS AGO AT THE age of nine i began reading. some of the many books written about pirates and shipwrecks by Edward-Rowe-Snow. My love for these subjects started two years before at the age of seven at the time two movies Titanic and Twenty Thousand Leagues Under the Sea, came out. The movies blew me away. After that i couldn't read enough books about maritime history. By age fourteen i had read eleven of Mr. Snow's books. I had to meet him. When we met a short time later, we hit it off right away. For we both loved the subject as well as history. Over the next twenty years, i was one of many who helped him with his lectures, book signing trips, and his tours of the boston harbor islands. All of the tunnels and dungeons that were normally locked, and closed to the public, were magically opened when he appeared.

We got quite dirty going down into the dirty cobwebbed tunnels, but, we took an extra change of clothes in case we got too dirty. It was a huge amount of fun. In his lifetime he wrote ninety-seven books and a small number of periodicals. He had started writing a volume on new england's greatest sea mystery--the loss of the paddle-wheel steamer *Portland* in november of 1898 with all hands. No one saw it go down in the huge one-hundred miles per hour gale that swallowed it up. Two days later much wreckage and bodies of the passengers and crew washed-up on the outer beaches of cape cod. Seeing that most of the wreckage was concentrated within a half-mile of beach in the town of Truro. The location of the wreck was thought to lie just offshore near the tip of Cape Cod off Provincetown. I was helping him with his research on his book but, the information trail ran dry in the nineteen-seventies. In nineteen-seventy eight, we were discussing the *Portland*, and he asked me that if i ever found the rest of the information on the *Portland*, would i please complete the book? Had i known it would take me thirty-two years, i might not have started it. But the nuns i had as teachers in my early years, taught me that your word must be kept no matter what, or your word is worthless. The nuns were some of the best teachers i ever had. When reading through his extensive research, and cross-checking facts, some of the reporters in the 1890's got it it all wrong, and most everyone took their word as gospel. One reported story from a schooner captain and his mate, said that they saw it wallowing in the heavy seas for two hours. As the eye of the fast moving storm passed over Provincetown, in which the sun came out and greatly improved visibility. How could a fast moving storm hover in one place for over two hours? I checked on this with the U.S. Weather Bureau in Taunton Mass, and they said no way. I think the the two sailors told the reporter a slightly different story, and it was changed and sensationalized to sell more papers. At the time of the sighting, the *Portland* was over twenty miles to the north in her death throes. This misquote in the papers threw everyone off in finding the wreck for ninety years. Through their twenty years of looking on their own showed the genius of both John fish and Arnie Carr and their helpers from American Underwater Search and Survey, *Portland* was finally found in 1989. They took photos of the wreck but could not prove conclusively that it was the *Portland*. In combining forces with the Jerry studds Stellwagen Bank National Marine Sanctuary, everything matched the historic parts of the ship. In september of 2003 and 2004 i got to go aboard the RV Connecticut to go to the wreck site. We had an rov five-hundred feet below on the wreck seeing things no-one had ever seen before--live. Mr snow's and my research combined into a total of 67 years. It's an eerie feeling being at the exact spot. I laid a wreath over the wreck in Mr. snow's memory and to the one-hundred-ninety-two passengers and crew that went down with the ship. It was extremely emotional for me, but exactly the right thing to do.

Table of Contents

Preface and Author Bio ·vii

Chapter 1	Building the Maine Masterpiece · · · · · · · · · · · · · · · · · · ·	1
Chapter 2	Take Her to Sea ·	8
Chapter 3	Cruising the Maine Coast ·	13
Chapter 4	A Child's Memories ·	17
Chapter 5	The Fateful Decision ·	23
Chapter 6	At Sea in Mortal Danger ·	32
Chapter 7	Joshua James Legendary Lifesaver · · · · · · · · · · · · · · ·	41
Chapter 8	Struggling to Survive ·	47
Chapter 9	The Dark Clouds Gather ·	53
Chapter 10	The Storm of the Century ·	59
Chapter 11	The Portland Inquiry ·	67
Chapter 12	The Storm Still Rages ·	75
Chapter 13	Filming the Sunken Portland · · · · · · · · · · · · · · · · · · ·	86
Chapter 14	New Developments ·	92
Chapter 15	Myths and Realities ·	95
Chapter 16	The *Portland* Victims – The Human Cost · · · · · · · · · · · ·	117
Chapter 17	The First Search for the *Portland* · · · · · · · · · · · · · · · · ·	119
Chapter 18	The New *Portland* Searches and Victim Stories · · · · · · · ·	123
Chapter 19	Expedition to the Portland, September 13-18, 2003 · · · · ·	127
Chapter 20	Side wheels verses Propellers · · · · · · · · · · · · · · · · · · ·	132

Chapter 21	Laying the Blame · 138
Chapter 22	The Sea Takes No Prisoners · 140
Chapter 23	Solving The One-Hundred-Year-Old Mystery · · · · · · · · · · 150

Epilogue · 159
Acknowledgements · 167
Bibliography · 179
Index · 183
Chapter Notes · 217
In Memoriam ·229

CHAPTER 1

Building the Maine Masterpiece

CAPTAIN HOLLIS BLANCHARD PEERED FORWARD into the darkness of the steadily mounting seas; he turned to his able helpers, Quartermaster Ansel Dyer and Second Pilot Lewis Nelson for advice. Should they keep sidestepping the steamer Portland further out to sea or work the ship back toward Gloucester Harbor with its many dangerous rocks and ledges?

Their total agreement seemed to be to hold the ship off the coast to ride out the seas in the forlorn hope that the immense storm now punishing the ship would subside by daybreak, giving them a better chance of making harbor with better visibility. The 280-foot side-wheel steamer had done this before when Captain Blanchard held the rank of First Pilot. These experienced seafarers had no reason to believe this would be any different. However, this was a New England coastal storm like no other. Before it was over, it would sink more ships and kill more mariners than any other storm in New England's history. Nature is absolute, and the power of the ocean has shown over and over again that no ship is unsinkable. As the seas grew steadily higher, the peril grew in intensity.

In early January 1889, at the shipyards of the New England Company in Bath Maine, carpenters began to set up the stem, or bow frame, for a very large steamboat. In the ensuing weeks, the stern, or back-end, frames would be constructed. The ship's keel, or backbone, was made of solid white oak, the hardest wood known at the time, would be laid between them. Fifty-foot lengths of this wood would be scarfed or overlapped, to form a keel about 250 feet long. The 30-foot long stern overhang would make up the remainder of the ship's 280-foot length. Shaped like the hull of a canoe in its underwater sections, it could slide through the water with less drag and keep its wake to a minimum.

Overseeing this work was William Pattee of Bath, Maine, one of the ablest ship designers of his day. Pattee had designed many kinds of sailing ships, schooners, steamers, and tugboats. Just prior to the *Portland's* launch on October 14, 1889, the Bath Daily Times described the vessel: "The frames are also of the same white oak with Hackmatack tops. Encircling the frames, just below the main deck beams is a wrought iron belt 7" wide and ¾ of an inch thick, secured to each frame by 3 four-inch screw bolts. From this belt, diagonal strips of 4 x ½ inch iron run.

"Up to the floor timbers, thus binding the frames into a solid body; the timbers are filled in solid beneath the machinery. The ceiling is of heavy yellow white pine, the bilge keelsons being of 14x15-inch timber, secured by four 1-inch bolts in each frame. The planking to four streaks above the load waterline is of selected white oak, the butts being secured by riveted composition bolts and the planking is of yellow pine. Above the boilers, there are iron deck beams, with a covering of sheet iron upon them to protect the deck planking above. The boat is divided by wrought iron partition into three watertight compartments, built according to the U.S. Government Steamboat laws. With adverse weather conditions possibly being an issue, one of the compartments could be filled with water and the vessel could still float. The main deck is of a 3-inch yellow pine. The cabins were to be finished in cherry, but the main stairway, leading up from the main deck, will be of mahogany. The main salon will be 225 feet long with a double row of staterooms on each side, the total number of these apartments being 167. The bulkheads between these rooms will be of diagonal strips, to add to the rigidity of the boat. On the hurricane or top deck, will be the quarters for the captain and officers, with 12 staterooms.

"There will be a double steering wheel, 6 feet in diameter, in the pilot-house, as the vessel will steer by hand, the smaller main wheel was powered to ease the job of the quartermaster, who steered the ship. The vessel will be lighted throughout with electricity. The engine will be of about 2000 hp with a 12-foot stroke, the paddlewheels being of 35 feet 10 inches in diameter.

She has a powerful steam windlass and will be provided with all modern fixtures of a first-class vessel. The boilers were made by the Bath Iron Works and are 23 feet 10 inches long, by 12 feet in height, and will weigh 60 tons each. They are two in number and built for a hydraulic pressure of 85 pounds, though they will work at about 53. On the outside of the paddle boxes is a representation of the globe, as on a map, the lines of latitude and longitude making up the open work. Above this, is a circle in which is the seal of the city of Portland. Above the pilothouse will be an eagle scaled up on a gilded globe.

The engines were made at the Portland Company's shops. As the Bath Iron Works were crowded with business, when the contract was taken, the new paddlewheel steamer will receive her boilers here at once, and is to be in Portland by November 1. She is to go on the line next May, to replace the John Brooks, which is to be taken off and sold. The commander of the vessel will be Captain William Snowman, and her first engineer, the able George H. Coyle. Ready for sea, she would represent $340,000."

The boilers and engines were installed after the launch to reduce strain on the hull of the ship as it slid down the ways. When the stern of the ship hits the water, it floats, while the bow is still on land.

This puts a tremendous bending strain on the midsection of the ship so that adding the heavy weights into the ship makes better sense when the ship is completely, supported by water, especially a wooden ship. The *Portland* was to be one of the most luxurious steamboats built up until this time, and was actually the zenith of Maine-built paddlewheel designs.

No expense was spared in its construction. Much expensive mahogany went into the paneling, stairways, railings, and furniture. Most of the seating in the main salons was covered with a beautiful red crushed velvet material.

Immediately after launch, the ship was towed by tugs downriver about a mile to the Bath Iron Works to have its boilers installed, this being accomplished by positioning it under a huge derrick, to handle the 60-ton weight of each. They were so large, that only two were necessary to power the 2,000-hp engine soon to be installed by the Portland Engine Company after a tow down to Portland.

The engine was of the old fashioned walking-beam type that had a 23-foot-long diamond-shaped beam that rocked back and forth to transfer the engine power from the single cylinder to a long crank rod to the paddle shaft connected to the paddle- wheels that actually moved the ship through the water. This ship would be one of the last wooden paddle wheelers built, as the design, even in 1890, was giving way to steel construction, which was stronger and to the screw propeller, which is far more efficient.

The rugged coast of Maine was one of the busiest steamboat routes with many companies and ships competing against the railroads for freight and customers. The steamboats would serve cities from Portland in the south part of Maine, to as far as Eastport, St. John's, and Halifax in Nova Scotia in the Canadian Maritime Provinces. Some of the lines also had routes far to the south serving Boston, New Bedford, and New York. These voyages could take up to three days one-way. The Cape Cod Canal would not be built until 1914.

The only way to go was the long way around Cape Cod, which was the scene of more shipwrecks per square mile than anywhere in New England. These routes were later extended as far south as Norfolk, Virginia.

The sides of the *Portland's* hull were of stout 3-inch thick white oak, while the bottom was of 4-inch thick southern yellow pine. This was even further stiffened by knee-shaped pieces of wood cut from the root of the tree to where the trunk starts vertically. Even with all this bracing, a wooden ship will still flex slightly, easing through the seas, absorbing shock and vibration, making the trip much more pleasant for the passengers.

Above the main deck was the superstructure that housed most of the passenger cabins, but it was constructed of the least noble wood. Hackmatack is a stringy and knotty wood that is practically rot proof. But it must be painted, as it cannot hold a stain. It is grown locally and is inexpensive to buy, and because it was installed on an area that houses a very long main salon, used mainly for dining and socializing, it did not affect the structural integrity of the ship.

Surrounded by a large number of cabins, the main salon was by far the most beautiful room on the ship, complemented by the grand staircases on both ends. A full-length raised roof covered this, according to the Bath Daily Times of July 1889: "The cabin finish for the main salon doors has molded panels on the inside while the reverse is in the Eastlake style. The main cabin will have a domed skylight while two octagon skylights will light the salon forward. Here there will be a gallery and two tiers of staterooms. Along the side there will be Corinthian pilasters with Corinthian capitals.

There will be 56 rooms on the hurricane deck and 168 below. Altogether, there will be 514 berths. The finish in the staterooms is cherry The berth in the gentlemen's cabin forward will all be protected by blinds, and will be hung with chintz curtains. The berths will be arranged in separate compartments, with three in each. The many maids and stewards in the crew pamper the passengers and make trips on the Portland Steam Packet line ships most enjoyable."

In a week and a half, the two huge boilers were installed downriver at the Bath Iron Works. The ship then began its journey to Portland under tow by a sturdy ocean-going tug. It arrived about eight to ten hours later at the outfitting dock, used by the renowned Portland Engine Company. The huge main cylinder, enclosing a single 62-inch piston whose up-and-down reciprocating motion generated the power for the huge 35-foot diameter paddlewheels. The piston travels a distance of 12 feet inside the cylinder, generating 2,000 horsepower. The *Portland's* was one of the biggest steamboat engines ever built by the company.

All of this was loaded into the ship over a three-month period, assembled, adjusted, and run up to break in the engine. During the breaking in, extra lines would be run over to the dock and the new engine would be run slowly at first, and then gradually increased in speed. With the twin side-by-side smokestacks in place, carpenters sent down by the yard enclose the top of the ship, which up until now had been left open and unfinished to facilitate the installation of the boilers and engine. With this accomplished, coal, anchors, and other safety gear were loaded aboard for the upcoming sea trials.

These trails would be supervised by the most senior captain on the line, Captain William Snowman, the son of another prominent Maine steamship captain, Thomas Snowman.

Between them, they had a total of 75 years at sea. That line would later extend to six generations to the present in the person of Captain Roger Snowman, the great-great-grandson of William, who comes from Weymouth and a friend of the authors for 30 years. His wife, Virginia, who is the family archivist, produced a photo of William Snowman that dates to the 1880s. At the same time, she produced another photo showing the crew, including

Captain William Snowman, standing on the forward part of the *Bay State* in 1895. A person would have to train for 10 years, first as a third pilot learning basic navigation, then on to second, then finally first pilot After serving in this important position for another three years, the first pilot could be considered to be promoted to the rank of captain.

CHAPTER 2

Take Her to Sea

IN APRIL 1890, THE LONG-AWAITED time to take the *Portland* to sea has arrived, lines are cast off and the ship begins its journey slowly down the harbor. Spectators lined the docks and shoreline to admire the beautiful ship, which is accompanied by a tug to safeguard on her maiden voyage. As she turned past the harbor forts and heads for the open sea, whistles and bells of the boats and ships in the harbor salute her. Dressed in white paint and trimmed with a band of gold, the *Portland* shone brightly in the daylight. Slowly at first, and then with gathering steam, she worked up to a cruising speed of about ten knots or twelve miles per hour.

Under the direction of Captain Snowman, the *Portland* performed beautifully. Along for this first trip are officials from the Portland Steam Packet Co., The US Department of Commerce, the Steamboat Inspection service, as well as the engineers from the Portland Engine Company who actually built the engine They checked the engine temperatures, steam pressure, and made sure the engine was broken in correctly. As the hours pass, speed was gradually increased to full power, usually held for three to four hours to test the ship's capabilities. At the end of the day, the ship returned to the Portland Engine Company's dock for further fine-tuning. Tests also are conducted to determine how much coal the ship burns at the various speeds; that is how mileage was calculated for a steamboat. Modification then would be made to the ship or engine as needed, or turned over to the government for use as a storage ship or a barracks.

A week later, all was well with the *Portland* and it returned under its own power to the shipyards to have the interiors finished.

On its return to the N.E. Shipbuilding yard, Superintendent Amos E. Haggert oversaw the final details. All told, it took 15 months to build the ship, which was completed in May 1890.

The following month, the Bath Daily Times reported: "The City of Portland is now complete and was opened for inspection today. From the excellent description of the Portland Argus, we glean the following factors of interest: William Pattee of Bath is the designer of the hull, which was built by the New England Co. of that city. The boilers were made by the Bath Iron Works and engine by the Portland Company. Her model is the ordinary side-wheel form, Convex sides, with long sharp bow and tapering Round stern after the yacht style. The vessel is painted white outside with the exception of the paddle Boxes. They resemble half of a globe; painted in colors and ornamented with a carving of the seal of the City of Portland at the top, which is gilded and colored in an artistic manner.

Watertight bulkheads separate the lower cabins, forward and aft of the engine. The forward cabin is used for a dining room. Both cabins are painted white with cherry trimmings. A tier of berths run alongside and are arranged in apartments with cherry-bound doors to protect the complex engine parts from the curious gaze of the passengers All of the stairways were made of mahogany with brass fittings.

From the forward salon by the richly curved stairways, one ascends to the upper saloon, out of which are alcoves leading into 42 cool and airy rooms.

The center of this salon is open to the main salon below, the opening protected by an elegant mahogany rail with curved balustrade, which adds greatly to the beauty of both salons, The officer's quarters are on the Hurricane deck. The main salon is about 220 feet long with a row of staterooms along each side. There are 168 rooms and 234 berths on the boat. The main salon is lighted with a domed skylight. It is furnished in the Corinthian styles of architecture and furnished with richly carved mahogany furniture with wine colored plush upholstery. The floors are covered with velvet carpets.

"On the main deck in the social hall is the purser's office. The hall is elegantly fitted up for a smoking room, with mahogany chairs upholstered in leather. Whist tables are convenient for those who delight in this pastime. Forward of the hall is the freight room, where 50 carloads may easily be stored. The sanitary arrangements were of the most improved pattern yet attained. The chief aim of the designers and builders has not been for beauty alone. Strength, safety, and convenience have been most important feature taken into consideration of all the comforts possible for passengers and the prompt dispatch of their large and constantly increasing business. She stands unsurpassed by any boat east of New York with the exception of the mammoth steamers of the Fall River Line, of which she presents a striking appearance."

Shortly after this, the *Portland* left the shipyard for its homeport of Portland to prepare for its first run to Boston. The ship had to be loaded with tons of coal, a nasty and dirty job but very necessary in the days before oil propulsion. In addition, the linens, bath supplies, freight, navigation instruments and the like all had to be loaded in preparation for the first paying passengers to come aboard. On a ship's maiden voyage, local dignitaries pull strings to secure a berth aboard the ship. This first run becomes quite a festive occasion, with the officers and crew in shiny new uniforms, and flags hung in the rigging and on the two masts.

The top officers of the Portland Steam Packet Co. also were aboard, including Mr. Coyle and Mr. Liscomb, the company's top managers and officers from the other Maine-based steamboat lines.

Crowds gathered at Franklin Wharf to see the *Portland* off on its first voyage to Boston. Its arrival at its destination was quite festive. Larger than any other of the paddle-wheelers of the Maine coast lines ever to dock in Boston, she presented a beautiful sight as it glided slowly into India Wharf on Boston's Atlantic Avenue. When coming into the dock, the engine was momentarily stopped, then immediately run backward for braking. The crew then threw small heaving lines to the dockworkers and the heavier lines were pulled across and made fast to the dock. The boat then moved very slowly ahead and swung into the dock using aft-facing spring lines.

The rest of the lines were then made fast to the dock and the gangplanks run across to the ship so the passengers and dignitaries can disembark. The whole party scene then happened again as all of the Boston dignitaries were shown throughout the ship by the captain and the chief engineer. After the tours of the ship are completed, the dignitaries then retired to a nearby restaurant for more celebration. The remaining crew restocked the ship for the return trip to Portland the next day At 280 feet overall, the *Portland* was the most luxurious ship ever to take the run down from Maine.

The design worked so well that in 1893, construction was started at the same New England Shipbuilding yard on a near sister-ship, the *Bay State*, which came into service in 1895, and worked the opposite run from the *Portland*. A "near sister" means that the ships look very much alike, but the newer one was slightly larger and longer, resulting from the slight advances in naval architecture at the time.

As the engines were broken in, the good weather passages could be made in slightly faster times, requiring about seven to eight hours for the 100-mile trip. When not in use during the daytime hours, the ship during the holidays would make special trips or excursions to locations up and down the coast, just as the commuter boats do today on summer weekends. This brought extra income to the company and helped fund the cost of newer ships.

Once a year, usually during the spring, the *Portland* would be hauled clear of the water into the dry dock. The bottom would be scraped and painted with the hull checked by an inspector from the Steamboat Inspection Service of the United States Government. If wood were found to be decayed or rotted, it would be immediately replaced and inspected again.

The annual haul-out inspection usually lasted from six to eight weeks included two or three of the top supervisors from the yard that built it and knew every plank and frame on the ship. In her later years, even more bracing was added so that the ship was made even stronger than when it was built. Borings into the frames and planks would also be made to check for internal rot in the wood. At the end of the docking session, the dock would be partly flooded for the planks in the hull to swell until any leakage was negligible. The ship would then be floated out and returned to Portland to resume its Boston schedule. The spare boat, usually the *Tremont*, would be put back into spare service or be chartered to another line in case of accident or overhaul of one of their boats. At least, this way, some money would be coming in, as fares had to be kept very low to compete with the railroads, which were taking more and more freight and passenger business away from the steamship lines.

Every few years, the Steamboat Inspection Service would require more safety gear, which made the boats more expensive to run. The only mishap in the *Portland's* eight years of service occurred on September 8, 1895, when backing out of India Wharf in Boston. It collided with the inbound steamer *Longfellow* coming in from Provincetown.

The *Longfellow* sustained most of the damage, with hardly any to the *Portland*, which continued to Maine shortly after the crew had checked it out. The steamship agents were left to take care of the paperwork. At this point, the Portland Steam Packet Company had two boats almost equal in size load on the run.

The two boats could now carry 800 passengers each and tons of general freight on the lower decks forward to make every trip as profitable as possible.

CHAPTER 3

Cruising the Maine Coast

—

SINCE THE TIME OF VIKING explorer Leif Ericson and *Mayflower* Captain Christopher Jones, attempting any kind of a landing on the coast of New England has been fraught with the utmost degree of danger. The coast is rocky and full of shoals that ships entering harbors during Northeast storms often were wrecked on an uncharted shoal or torn into pieces on a very rocky shore. Luckily, Captain Jones of the *Mayflower* found the tip of Cape Cod during good visibility and anchored the ship in Provincetown Harbor, instead of smashing to pieces on the outer shoals or beaches.

New England weather has its origins thousands of miles away in the arctic reaches of western Canada, as well as the Gulf of Mexico. As the rotation of the earth is from west to east, the weather fronts usually come from the northwest Pushing this weather along at high speeds several miles up is the jet stream. This separates the cold air in the north from the hot air in the south. However, at times it undulates in a wavy pattern, going from north to south. During these undulations, hot and cold fronts collide, causing fierce, raging storms accompanied by a moving low-pressure system that can suck in energy from miles around. When a system such as this combines with another low front, the result can be cataclysmic. When it goes over water, it picks up much more energy from the ocean moisture, often with winds reaching more than 100 miles per hour. This storm swirls around a center point and actually develops an eye just like a hurricane, doing even more damage. At sea, it jams the waves so close together that they have nowhere to go but up.

During sudden increases of the storm's power, it can generate rogue waves 100 feet high. These waves have caused ships and fishing boats to disappear from the face of the earth with no time to radio a distress signal. The propellers or paddlewheels of a ship do not work well in this environment.

Off the coast of New England lie two ocean currents. The cold one is called the Labrador Current, as it naturally comes down from the Bay of Labrador, near Sable Island off Newfoundland. It collides with the northerly flowing Gulf Stream that consists of warm water from the Gulf of Mexico, hence the name Gulf Stream. With no wind present, this very often results in thick fog with visibility less than 50 feet. In past centuries, fishermen and seafarers have reported seeing shapes resembling "ghost ships" in the fog, a part of the lore of the North Atlantic.

If a storm rotating clockwise forms in this location, the storm will gradually move out to sea. However, if it is rotating counterclockwise, it will likely wind back toward the coast of New England, lashing the area with seas of up to 40 feet. This pushes a tidal surge that can cause coastal flooding with a surge of 10 feet above the normal high tide level.

With the jet stream above the earth moving at more than 100 knots, any changes in its north-south separation can slam both the cold and warm moisture fronts at great speed to form an immediate storm.

As the storm picks up moisture, for example, over the Great Lakes, it gains strength over the water and loses strength while crossing land. When the jet stream bends to the south, it forms a trough. If it bends to the north, it is called a ridge, often mixing the hot air from the south with the cold from the north. When these are kept separated, the weather is usually fair. In the winter, at about 0 degrees, during a storm at sea or on land, the snow falls almost horizontally, reducing visibility to almost zero. Trying to keep a ship at sea pointed into the wind becomes next to impossible. To add to all this, the salt spray turns to ice all over the wheelhouse windows, which must be cleared just to see. Usually the front center window must be kept open just to see ahead.

The ice adds to the ship's weight above the waterline, greatly increasing the risk of capsizing. This also causes the ship to roll much more than normal, adding to the seasickness of the unfortunate souls aboard. If the ship goes down and crewmembers and passengers manage to get off, even with a lifejacket on, a human can only survive about a half an hour in the frigid water before hypothermia sets in. The cold dark Atlantic Ocean does not form any distinction of what it takes; all souls are in equal peril. Even today, there is no such thing as an unsinkable ship, even though life can be made more comfortable. Nature is still absolute.

Seas this huge can blow houses right off their foundations and smash the wood into matchsticks. For ships caught at sea, the result is oftentimes fatal. After a really severe cyclone, wreckage covers the beaches from Maine to Delaware. The very worst of these occurred on November 26 and 27 in 1898. It became nationally famous as the "*Portland* Storm," named for the steamer whose loss of 192 souls represented the largest loss of life on a single ship.

In all, more than 450 sailors lost their lives. It also sank 150 ships. Damage up and down the shore would total in the billions in today's dollars. From Boston to Cape Cod, electric power and telephone lines were knocked out for weeks The telegraph cable to Europe also was severed; the broken ends had to be fished out of the water off Cape Cod and repaired at sea. Huge snowdrifts blocked railroad transportation to Provincetown, and downed utility poles that took over a week for repair crews to clear.

The first captain of the *Portland* was William Snowman, who was born in Penobscot on September 21, 1830, the son of Captain Thomas and Sarah Snowman.

"At an early age, his family moved to Newburyport, MA. where he received his education, graduating high school at the age of 16," The Portland Press article of April 5, 1898 reported. "He was associated with his father in the coasting trade for three years and when at 21 years of age, he commanded a coasting vessel himself. He was 12 years in this business chiefly between New England ports, and the last three years of his connection with the coasting trade, commanded a packet sailing between Portland and Boston."

In the spring of 1863, he joined the Portland Steam Packet Company as second pilot, a position for which he was well qualified, as he knew every turn and shoal of the New England coast. He was second pilot for 10 years, and then was promoted to first pilot. When Captain John Liscomb died in 1881, Captain Snowman was made captain of the *John Brooks*.

He had been in charge of every boat of the line at different times except the *Tremont*. From the *John Brooks*, he succeeded to the command of the *Portland*, later being in charge of the elegant *Bay State*. Snowman was the oldest captain in service during his employment of the steamship company, and during those 30 years, he was off duty only six weeks. From his perfect knowledge of the coast and his vigilance, he never met with an accident; his long records were remarkably clean."

CHAPTER 4

A Child's Memories

EDWARD ROWE SNOW, THE WELL-KNOWN author of sea stories, often received information on historic events from readers, such as this description of the *Portland* from a 1965 letter from John Poland of Camden, Maine:

"*Dear Ed;*
I am interested to learn that you are engaged in a book that is chiefly about that gloriously beautiful boat, the SS Portland, of the old Portland Steam Packet Company of so very many years ago. It was one of my favorite ships as a boy. I have pictures of it gleaned from this and that travel gazetteer of those early years, back in the 1890s. And, I well remember that awful day when news came that the Portland was lost; it was as though some old friend died.

My folks came from Maine and we spent our vacation days down in Boston. Sometimes we went by train, taking the cars at the old RR station in Haymarket Square; but equally often took the Portland boat from India Wharf. We usually took the little open car electrics from Park Square. We had gone in on the 5:20 train from West Roxbury Station and jogged our way to Atlantic Avenue. Sometimes when I was with my older brother, we walked the distance across Boston and on route passed down through Broad Street where, to this day, over seventy years later, in memory, I smelled the coffee being ground and bagged. And, as we approached the wharf, we saw the big steamer backed up toward the avenue *and wearing a large canvass sign hanging down from the top deck bearing the information "This steamer sails tonight at 7 o'clock for Portland. Fare: $1.00.*

It was always a joy going down on the boat. To my childish eyes the ship was enormous! And yet, so was our horse, Dan, a big animal. When I learned that Dan was going onto the boat, too, I was a bit fearful lest he be too much for even those vessels.

The very gangplank leading into the fair palace was a promise of greater pleasures to come. And the insides with the grand staircases, the carpeted floors, the plush chairs and larger seats, the open forward salon with its gallery above, and the staterooms with those interesting bunks, one smaller over a larger beneath, and the people coming and going; all these experiences for a boy from four to twelve years of age were fore-tastes of a sort of heaven. For, I can't remember when I actually began to sail on the Portland, but I was over twelve when she sailed out on her last voyage. I suspect I was carried aboard as a babe-in-arms at first.

But I do remember this; I must have just passed my sixth birthday was very proud and conscious of my increasing years. I was with my mother out in the salon and there were other passengers and friends moving about, and a ship's officer the purser most likely came around, threading his way between folks and collecting the fares. As I remember back now, I wonder why that was the way it was done instead of, as later, getting the tickets at the gangway the next morning. Anyway, this officer reached us and began taking our tickets as he looked inquiringly in my direction to learn if I was still of the free rider age (under six)? I had been ill; I was small for my age; and I could have well have passed for five.

My mother doubtless wished to save a cent and declare my status and assured the man that he is "five". But I was not to be insulted that way and I spoke up and the officer smiled knowingly; and I guess my mother *had to pay. But, after all, it could not have been more than 50 cents. Well, maybe, we were poor in those days and had to pinch pennies.*

And I remember this too; as we made our way about, there in the big salon, we had to walk around mattresses being laid out on the floor. I can still see folks standing by the allotted mattresses, perhaps lying on them. So great was the demand for space in those busy years.

Once in our own stateroom, mother always rang the bell for the stewardess to fetch us a pitcher of ice water, as we brought our own supper. I can't remember ever having gone into the dining saloon. And there seated on one stool or on the end of the berths, we munched our sandwiches and had our cakes and drank the ice water. But that was like a picnic. And what boy ever scorned a picnic! And mother and Bertha, and Etta, and Willard made it all a bit of home right there in our cabin. It was a lark! But the hour passed; and as we sat there eating suddenly someone said, "Look! The wharf is sliding backward!" And sure enough, it was slowly moving along and with definitely gathering speed. And then the great and deep-throated whistle boomed somewhere overhead to announce to all in front of the wharfs slip that the steamer on going forth, and to watch out. And we could now hear the slap, slap, and slap of the big paddles of the wheel. No boy is ever satisfied to remain cooped up; no matter how glamorous was the stateroom neither at first, nor for that matter the grand salon either. The open deck was the goal."

The trip from Boston would usually take between seven and eight hours, depending on weather and sea conditions. The ship would leave Boston at the usual sailing time of 7 p.m. and get into Portland at about 3 a.m. This is why the steamboats came to be called the "night boats." The terminal, at Portland's Franklin Wharf, was very close to the terminal for the Grand-Trunk Railway to other parts of Maine, so an easy connection could be made from one to the other. A similar sailing would leave New York City as well. If stormy sea conditions happened along the voyage, it would take two to three hours longer. Fog was another factor that could delay a trip, as the boats would take a longer course farther out to sea to avoid hitting rocks and reefs that were close to land.

The captains and pilots would listen for the distinctive sound of different whistles and fog signals from the buoys and lighthouses along the way as they groped their way through the pea-soup fog. The safety of the passengers and crew were always first, as the Portland Steamship Line, started in 1844, and had never lost a passenger in over a million miles of sea travel.

When approaching Portland in fog, the pilots and captain would come in at a slightly right angle to land while listening for the fog and whistle signals of the Old Anthony Ledge Buoy or the Portland Lightship. They would then set a course to pass by Portland Head–light, which marks the entrance to Portland Harbor. From there, it is a simple navigational exercise to get to the inner harbor to tie up and discharge the passengers, as well as prepare for the next voyage.

After loading coal and provisions, the ship's crew could then return to their homes until the next day's sailing. If during a voyage, a bad storm sprang up, the ship could duck into the nearest suitable port. If this were impossible, the captain would head offshore for some maneuvering room, while keeping the vessel's bow into the wind and waves to ride it out safely. A sidewheeler in these conditions tends to take a battering under the overhanging guards forward of the paddlewheels, so in extreme conditions, cargo can be shifted to one side of the ship to raise the damaged side higher out of the water to lessen the danger until permanent repairs can be made in a shipyard. If this were the case, the Portland Lines, 'the spare boat, the *Tremont*, would be put on the run in its place until the repairs were finished.

Once each season, usually in the spring, the steamboats would be taken out of service for its annual overhaul for a period of six oi eight weeks. During that time, the wood hull would be allowed to partly dry out so the bottom could be scraped clean, repainted with anti-fouling paint, and thoroughly checked for rot. The bottom would be painted with antifouling paint to keep grass and sea creatures from attaching themselves to the bottom slowing down the ship If any rot was found, it would be replaced with new planking and an inspector for the U.S. Steamboat Service would check and sign off on the final repair work. All ships inspected for safety by the government would be issued a new annual operating certificate.

During these annual overhauls, extra cabins would be added or deleted for the convenience of the passengers and crew. In addition to passengers, the ship would also carry general freight between ports. The freight would be stowed in the lower forward part of the ship, as this would be the least desirable part of the ship for the passengers. However, it would be home for the crewmembers, their quarters were quite meager compared to the luxurious accommodations of the passengers. All in all, the crew's quarters were still better than the caboose that a train crew would get; the food was much better, and the ride was much smoother than on the train, which was the main competition for the steamboats. In the winter, when severe cold and freezing temperatures would cause rivers to freeze over, the boats would not run at all.

In the 1890s, as there were no icebreakers, snowdrifts would cause the trains to grind to a halt. Without cars as part of everyday life, most travelers at that time used the boats or the railways. It was an all-day affair to travel any distance. Add to this many bulky steamer trunks and fidgety children; it could make any trip exhausting.

But passengers on the *Portland*, like Maine's John Poland, thought it was a joy to travel on the *Portland* and her near-sister ship, the *Bay State*, because of all the goings on during the departure and docking – the blowing off of steam pressure, the wonderful sounds of engine room bells and gongs sounding their raucous tones and chimes, and the wonderful sight of the main walking-beam of the huge engine going up and down constantly.

After a while, it could be mesmerizing to watch, while in the beauty of the grand salon, or the main room of the ship, the activities of the passengers would be cause for much people watching. Meals would be served onboard for an extra price, or passengers could bring their meals. The crewmembers would be smartly dressed in their crisp uniforms, catering to all the passengers' needs.

The salt air, the gentle rocking of the ship and the sound of the paddlewheels would all serve to put one to sleep. This had a very soothing effect on the children, giving the parents some respite.

With all the risks inherent in coastwise travel, collisions and grounding were very infrequent on the Portland line.

CHAPTER 5

The Fateful Decision

DURING THE 1890S THE PORTLAND Steamship Line, in business since 1844, had steamed more than a million miles without an accident other than an occasional grounding, and also had never lost a passenger. This could lead to a false sense of security.

The *Portland's* near-sister, *Bay State,* was the second last to be built with massive radial-style paddlewheels in 1895. A newer and more efficient feathering-type paddle wheel had been in use for some time. The old style radial wheels resulted in wasted power. The more open water the boats were designed to run in, the deeper the floats on the paddlewheels would need to be set. Also to be accounted for was the rocking of the ship from side to side. The newer, feathering wheels had a more complicated mechanism that required more maintenance but provided more efficiency and less drag, leading to more speed and less coal usage.

Also, with smaller wheels, the huge paddle boxes could be dispensed with, resulting in cleaner lines and more cabins. The number of buckets on the paddlewheels and their spacing had to be matched to the very slow turning speed of the walking beam engines of only 17 to 25 revolutions per minute. In the 1800's, folks going out on a daily fishing trip would anchor their boat in the middle of the busy channel in the fog and fish. In the days long before radar, steamboat captains had to listen for fog signals from lighthouses, which all had their own distinctive sounds. A fisherman in the middle of the channel would suddenly hear the swishing sound of a paddlewheel boat coming out of the fog and think nothing of it. He was unaware of the mortal danger that was coming. If the steamship did not see him in time, and turn to avoid him, the boat would be sucked into the paddlewheels and chopped to pieces.

This never happened to the Portland Steamship Line, but it did happen on other lines. If a passenger was sitting on the rail forward of the paddlewheels and fell overboard, he or she would be ground up in the paddlewheels, which could never be stopped in time as they were directly coupled to the engine. Some of the old steamboat captains would say that hours and hours of boredom are mixed with moments of terror. Each time a steamboat would dock, as the engines were not reversible with a gearbox, the engine would have to be stopped short of or past top-dead-center, or the engine would become hung up. Eventually, a timber would then have to be inserted on the inside of the paddle box by removing a panel and prying the wheel by hand. Large levers would close the intake valves and other valves would be made to open, causing the engine to run backward.

By that time, a collision would have taken place. It had to be done this way to stop the beam-engine steamboat, as they do not have brakes. The only way to stop the steamboat was to reverse the engine after stopping. This made docking an interesting maneuver, to say the least. The Portland Line's safety record and many other factors led up to the loss of the *Portland* with her crew and passengers.

Captain Deering of the Portland Line retired in the early 1890s, and became its manager. In April 1898, the senior captain of the line and the first captain, William Snowman, suffered a stroke at the wheel of the *Bay State* on its way to Boston. He died within minutes. In the ensuing months, some of the other captains of the Portland Steamship Line took over the two ships.

In mid-November 1898, the senior First Pilot, Hollis Blanchard, was promoted to captain and assumed command of the *Portland*. Days later, the next senior pilot, Alexander Dennison, also was promoted to captain and assumed command of the *Bay State* The two steamboats ran opposite one another on the Portland-Boston route. Sometimes, when both boats were in port together, a change of command would take place, with Blanchard taking over the Bay State.

The recent death of John Coyle Jr., the Portland Steamship Company's general manager, prompted the company to promote Boston Agent John Liscomb to the position, even though he had not been a captain of one of the steamboats. The board of directors of the steamship company felt he was qualified for the position.

With two new captains skippering the ship's, resources were by now starting to be stretched a bit thin. Before leaving Portland for Boston on Thanksgiving Day, 1898, First Mate Edward Deering had overseen the loading of 90 tons of coal aboard the ship. These were divided between the main and auxiliary coal bunkers with about seven tons' worth in sacks on deck. As Deering would not be about for the trip back, he tried to cover every eventuality that could be foreseen in this regard. Recently retired Captain Charles Deering had died on Thanksgiving at his home in East Boston, and Captain Blanchard had given permission for his first mate and First Pilot Louis Strout and Purser's Mate J.F. Hunt to miss their return trip to attend Captain Deering's funeral.

As the ship was three officers short, there would be a problem if the lifeboats had to be launched. The lifeboats were of a new design, built out of steel, with sealed floatation chambers at each end. The Thanksgiving Day journey to Boston was uneventful, with the *Portland* backing in and tied to the south side of India Wharf, facing the main channel.

With its sterling reputation for comfort and safety and its yacht-like profile with its side-by-side twin stacks, it looked powerful and graceful. Its twin masts were rigged to carry a sail if an engine breakdown occurred. This was far from likely, due to the reputation for reliability of the Portland Company's engines.

The time at India Wharf would be taken up cleaning the cabins, as well as loading freight in the forward part of the ship. One hundred tons of general freight was being shipped to Maine stores for the Christmas sales that would be in full swing in a short time.

Saturday, November 26, 1898, started out sunny but cold. In the early afternoon, the skies started to cloud up with a slight westerly breeze. In a typical Thanksgiving weekend, the boat lines would be booked with many passengers returning home to Maine after visiting with relatives in Greater Boston. Later in the afternoon, the sky turned bronze, which sometimes appears before a major storm. As was his custom, the skipper went to the office of the National Weather Bureau in downtown Boston to check the latest weather maps. Two storms were brewing in the Great Lakes and the Gulf of Mexico, heading eastward.

These two low pressure fronts later combined off the North Carolina Coast, growing stronger by picking up the moisture from the ocean. Normally, fronts like this would spin off into the Middle Atlantic, and then head for Bermuda or to Iceland. However, the counterclockwise rotation of the storm would pull it back in toward the coast. Weather advisory called for heavy snow in the New England area, starting Saturday, lasting into Sunday. It was snowing in New York City, with the wind backing to the Northwest. This would be interpreted to mean that the storm would pass to the south of New England and the *Portland* would make port ahead of it, as the captain had done many times before. Shortly before returning to the ship, he stopped at the Portland company offices. Agent Williams told him that General Manager Liscomb had called from Maine, suggesting that Captain Blanchard delay the ship's departure until 9 p.m. to gauge the weather better. If it was quite bad, Captain Blanchard should cancel the night's trip all together. Blanchard promptly called the company's offices in Maine to speak with Liscomb, but the general manager had already left by train for Boston to attend Captain Deering's funeral. Instead, Blanchard spoke with the newly appointed Captain Alexander Dennison of the Bay State, who told him that he was following manager Liscomb's suggestion to wait until 9 p.m. The captains were bound to follow orders only; suggestions were left up to their discretion.

After returning to the ship, Blanchard was visited by his oldest son, Charles, who lived in Boston. They discussed the approaching storm, but Blanchard assured his son that the storm would pass to the south, and then out to sea, as the wind in New York had veered to the northwest. Before the start of the voyage, Blanchard had also talked with Carrie Courtney, formerly of Lewiston, Maine, who was a passenger on the *Portland* that night. In a letter to author Edward Rowe Snow, she said Blanchard had shown her a watch for a gift for his daughter's coming-out party the next week. Courtney had changed her mind and took the train instead.

Over the years, when Mr. Snow would give a lecture on the *Portland*, someone would invariably tell him that a relative was supposed to have sailed on the ship. However, as this person had gotten very inebriated at a local saloon, he missed his voyage on the *Portland* that night. An estimated 8,000 people have said that their loved ones had missed their boat that fateful Thanksgiving evening. If everyone had made it to the steamer, the disaster would have happened right at India Wharf, as the ship would have sunk at the dock under the intense weight of more than 8,000 passengers. The ship had a capacity of 800 passengers at most.

At precisely 7 p.m., a prolonged blast from the *Portland's* steam whistle announced that the ship had slipped her lines from the pier and was slowly gathering way, as the giant paddlewheels thrashed the water of Boston Harbor. Snow had not yet started to fall as the *Portland* cleared the pier. As her speed increased, she swung onto the channel and pointed her sharp bow toward the lower harbor where she would take a slight turn to port, and head out into the open sea as she had many times before. With a draft of only 12 feet, she could sail out thru the not-yet-dredged broad sound passage and head directly for Thatcher's Island off Rockport, taking another slight turn to port and steam directly for Maine to pick up the next marker at the Isle of Shoals close to New Hampshire.

At 7:37 p.m., as the *Portland* was passing Deer Island Light, at the lower end of the harbor, light snow had begun to fall. Wesley Pingree, the lighthouse keeper, in a later interview with Mr.

Snow thought that it had begun snowing at 5 p.m. However, the weather service said that the time was 7:37 p.m. It was commonly believed that the *Portland* was never seen again, but nothing could be farther from the truth. Captain Collins of the steamboat *Bangor* had started out, but had decided to turn back, anchoring in the President Roads, in the lower part of the harbor.

As the *Portland* steamed by shortly afterward, he signaled her by whistle, but the *Portland* headed on out. About a half an hour later, according to testimony following the disaster, Captain William Roix of the incoming steamer *Mt. Desert* passed the *Portland* abreast of Graves Ledge, outside the harbor, expecting the *Portland* to turn around and head back. However, the *Portland* kept on going into the night's gloom. Upon nearing Gloucester, the *Portland* was sighted by Captain William Thomas aboard the fishing schooner, *Maud S.*, three quarters of a mile southwest of Thatcher's Island. Upon reaching the island, Captain Blanchard decided to run very close to the shore between Thatcher's Island and Londoner Ledge, about five hundred yards from the twin lights.

At this time, 9:30 p.m., the seas were normal. In a letter to Mr. Snow many years later, Captain A.A. Tarr of Thatcher's Island said the weather was so normal that he barely bothered to look out to see the steamer's lights. At this time, Captain Lynes B. Hathaway of Brockton, a master workman for the U.S. Lighthouse Service, observed the lights of the *Portland* as she passed a mere 500 feet from the shore. The *Portland* then continued its voyage up the coast. Ninety minutes later, at 11 p.m., Captain Reuben Cameron of the schooner *Grayling* sighted the *Portland* 12 miles southeast of Thatcher's Island. The *Portland* by now had made a radical change of course. It came near colliding with the *Grayling*, but Captain Cameron warned her away by lighting a Costin Flare to avoid a collision. Cameron noticed that the *Portland* was swaying and heaving badly, but saw no damage.

At 11:15 p.m., Captain Frank Stearns of the schooner *Florence Stearns* saw the *Portland*; a half an hour later at 11:45pm, the *Portland* was seen bearing down on the schooner *Edgar Randall*, which was 14 miles east/southeast of Eastern Point, at the head of Gloucester Harbor. Captain D.J. Pellier rapidly changed course to avoid being run down. At this time, the superstructure of the *Portland* appeared to be damaged, which could account for the *Portland*'s loss of control. If any of the ship's steering gear had broken down prior to this, it would account for the several near collisions. The storm was now greatly increasing in strength and intensity. Meanwhile, at approximately 8 p.m., Captain Albert Bragg of the steel steamer *Horatio Hall* left the dock in Portland Harbor for his cruise to New York City. Being a propeller steamer with a high bow and no overhanging guards to get slammed by the seas, she cleared Portland Harbor and headed south. His course to New York City would take him directly south, 16 miles east of Boon Island and 40 miles east of Thatcher's Island.

Upon reaching a position 40 miles east of Thatcher's, the *Horatio Hall* was hit broadside by the storm with such savage fury that Captain Bragg was obliged to turn the ship into the wind to ride it out, much like what the *Portland* was doing. "I never saw such a storm in my life" Captain Bragg said later at the inquiry of the *Portland*'s loss.

The *Horatio Hall* was a steel ship, much stiffer than the *Portland*. All that any ship can do when getting pounded by 40-foot waves is to turn into it to present the smallest target for the massive waves. Even with a deeply located propeller, in seas that huge, the propeller would be coming out of the water some of the time. Special precautions had to be taken by the chief engineer to back off the throttles with the propeller churning air, not to over-rev the engine and break a critical part at the worst time. Conversely, when propellers thunder deeper and deeper under water, vibration occurs, so the main shaft bearing has to have extra lubrication by a fireman stationed at each bearing, pouring extra oil on them.

On early engines, the flywheel and main crankshaft were exposed with the connecting rods, throwing oil all over the place. In this grave situation, the engine room crew would have to wear oilskins or raingear to keep from becoming oil soaked, even in a 100-degree engine space.

During the next five hours, the *Horatio Hall* was pushed all the way across Massachusetts Bay and was off the tip of Cape Cod the next morning. With 40-foot waves and driving snow, visibility was, at best, 200 yards. At other times during the gale, visibility dropped to zero. Furthermore, off Provincetown that Sunday morning was the 128-foot propeller steamer *Pentagoet*, steaming from New York to Rockland, Maine, under the command of Captain Orris Ingraham with 18 passengers and crew. It was fighting for its life as well.

Ashore at Rockport, Mass., Captain Frank Scripture said that "the storm seemed to travel in veins, in which it blew down rows of trees and barely touched others." At 5:45 a.m., Captain Samuel Fisher of the Race Point Lifesaving Station at the tip of Cape Cod distinctly heard four short blasts from a steamer's whistle while standing at the foot of his bed. He immediately went into another room and looked at the clock, then rang the gong signal for the lifeboat. Immediately after, he telephoned the nearby Peaked Hill Bars Station, and ordered a surf man to go down and check the beach for any incoming wreckage. This storm was the worst Captain Fisher had ever experienced.

The storm continued pounding the Cape until 9 a.m. Sunday. Just then, an astounding thing took place, as Captain Benjamin, the Cape Cod District Superintendent later described" "Between 9 and 10 a.m. Sunday, November

27, there was a partial breaking up of the gale, the wind became moderate, the sun shone for a short time and the atmosphere cleared to an extent which disclosed two coastwise steamers passing southward, also a small fishing schooner lying to under short sail. About 11 a.m., the gale again increased and vision was obscured until after daylight Monday morning".

As described, the captain's experience was remarkably similar to the eye of a hurricane passing overhead.

CHAPTER 6

At Sea in Mortal Danger

—

IN A 1940S INTERVIEW WITH Edward Rowe Snow, Captain Charles Martell of the Boston tugboat *Channing* reported the tug passing close by the *Portland* on its way out past Nahant.

"I was steering in a southeasterly direction. We were off of Nahant. The weather was not bad at the time, but I knew a serious storm was coming. There were 10 to 12 young men gathered on the topside of the *Portland*, just forward and aft of the paddlewheel box on the port side of the ship. When one of the Youngblood's on the *Portland* shouted to me to get my old scow out of the way, I shouted back, 'You'd better stop that hollering, because I don't think you'll be this smart tomorrow morning.' By this time, I was less than 20 feet from the *Portland* and could easily make out the features of the young men sailing to their death. I gave three blasts of the Channing's whistle, and Captain Blanchard, whom I could easily recognize in the wheelhouse, answered back."

By this time, however, the two low fronts had combined off the coast of North Carolina in the area of Cape Hatteras (known as the graveyard of the Atlantic), and headed northeast at high speed, gaining colossal amounts of strength as it passed over the ocean. A huge counterclockwise, or anticyclone, had formed, with a size approaching 600 miles in diameter. Nature set in motion the fiercest storm ever to hit the New England Coast. In late November, in the New England area, the typical nighttime temperature is usually in the high teens to low twenties, well below freezing. The temperature of the ocean would be about 26 degrees, and the only thing that keeps it from freezing over is that it is constantly moving.

When a fierce cyclone with winds approaching 100 miles per hour churns up the ocean, waves can reach heights from 60 to 100 feet high, with occasional rogue waves even higher. These rogue waves strike without warning and have, over the years, been responsible for boats and ships disappearing without a trace. Not even in modern days does a ship have time to even get out an emergency "mayday" call before being swallowed by the ocean. In the October, no-name gale of 1991, the seas swallowed the 65-foot fishing trawler *Andrea Gail* almost without a trace. All that was found later were some barrels with the letters AG on them and an emergency position radio Beacon that had washed up in the off position on Sable Island off Newfoundland. Even with a length of 280 feet, the *Portland*, or any other vessel, could fall victim to such a fearsome wave. When the killer winds of a cyclone push that much water, waves get squeezed together and have nowhere to go but up. This causes the sea to smash anyone or anything in its path.

During a storm at sea in 1910, the 785-foot-long Cunard liner *Lusitania* smashed into a wave that flooded the bridge 60 feet up to a depth of five feet, knocked loose some of the engine telegraphs that had been bolted to the deck, and pushed the front wall of the bridge back five feet. A similar thing happened to the Cunard liner QEII in the 1970s when an 80-foot rogue wave hit the bridge over 80 feet above the water, damaging the ship and slightly injuring several passengers. As Dr. Robert Ballard once said, "Murphy of Murphy's Law lives at sea." The hostile environment multiplies anything that goes wrong at sea.

If the steamer *Portland* became disabled and swung crosswise to the huge seas, the waves would roll it over and capsize the ship. To get more power out of the engine the engineer could tie down the sickles steam pressure blow-off valves to build up incredible pressure, getting more horsepower from the engine. An engineer could lose his license for this, but in an extreme emergency, desperate measures must be taken. The trick is not to raise the pressure so much as to cause the boiler to explode, leaving the ship helpless and at the mercy of the sea. If the coal in the bunkers began to run low, more had been stored in the sacks on the main deck, but if these were exhausted, the ship could be kept going by using the furniture, burnable cargo, and parts of the structure itself.

The steamboat *Katahdin* had successfully done this during a tremendous gale in January 1886 at Cape Porpoise in Maine. She fought through huge seas for more than 10 hours through awesome struggles of her officers and crew. The battering she took put her remaining coal supply under seven feet of water. Furniture, cargo, and wood from the ship were used as fuel. The Katahdin limped into Portsmouth, N.H, for respite and temporary repairs. She was refueled in the interim and headed on to Boston after the storm subsided. Everything that was stored on the deck had been swept completely clear. Being larger and more robust than the Katahdin, the *Portland* would be expected to ride out rough seas more easily.

During storm conditions, ships tend to burn more fuel, as the throttle is set at a much faster speed. The ships do this just to keep up with the seas. With the necessary constant maneuvering of the ship with the amount of fuel aboard, parts of it being underwater in a partly flooded bilge, she would be expected to be kept running for about 27 hours. The *Portland* also carried two masts with working sails in case of an accident with the machinery. Even though steam engines had been mostly perfected over the ensuing years, passengers would be ordered by the ship's crew to don the cork filled life-jackets as a safety precaution. The passengers also would stay in their cabins, out of the way of the crew.

Probably, the lights failed sometime between 11 and 12 o'clock. The darkness intensified the strange sounds of the ship against the background of howling winds and pounding seas. In such circumstances, it would be easy to expect that each moment might be the last. From midnight to the sunrise hour, barometers on the tip of Cape Cod registered 28.80 inches.

Early Sunday morning Captain Samuel O. Fisher of the Race Point Lifesaving station heard a steamer's whistle blowing four short blasts, which is a distress signal, but it was promptly drowned in the roar of the wind and sea. Captain Fisher sent his man out along the beach, but nothing more was heard and nothing could be seen through the thick driving snow."

A short time later, Captain Michael Hogan of the Schooner *Ruth M. Martin*, was struggling through the storm to clear the dreaded Peaked Hill Bar and enter Provincetown Harbor on the back side of the Cape. Being in a position four miles off Highland Light, in the town of Truro, he sighted two ships nearby. One was thought to have been the *Portland*, as it looked like a paddle wheeler in the distance. The other ship could have possibly been the smaller 128-foot screw steamer, *Pentagoet*.

Captain Hogan then raised a flag of distress by flying the American flag upside down hoping one of the steamers would come to his aid. At this time, the center, or dry hole, of the storm passed over the tip of Cape Cod, bringing subsiding seas, with the sun showing through slightly. This was only a short respite of an hour at most, as the storm would immediately worsen from a slightly different direction. The four short blasts mostly likely came from the *Pentagoe*t, which was also fighting for its very survival. The *Portland* was not the only ship to sail from Boston that fateful day.

At 4:15 p.m. the Merchants and Miners Line steel steamship *Gloucester*, under the gifted command of Captain Francis M. Howes, sailed down Boston Harbor on its run to Norfolk, Virginia. The *Gloucester* had been fully loaded with freight, which put the heaviest weights way down in the bottom of the cargo holds, stabilizing the ship with a very low center of gravity.

Captain Howes was one of the ablest skippers on the Atlantic Coast. In the days without buoys to mark all of the hazards, courses at sea had to be judged by the method of dead reckoning, which is still used today. According to Captain Howes's account to Edward Rowe Snow, "At 7:30 p.m., the *Gloucester* was heading south off Cape Cod, with the wind blowing 70 miles per hour. He ran the ship at the maximum speed possible so that if he struck shore, the ship would come high on the beach so passengers and crew could get off safely.

Usually the ship could do 15 knots of speed, but thanks to fine-tuning adjustments to the engines and the superhuman efforts of the engine room crew, the *Gloucester* was able to achieve a fantastic 18 knots, a remarkable speed for ocean steamships of that time. Gradually, snow began to fall and visibility was reduced, at times, to less than 50 feet. He calculated this by the forward lookout, which was only 50 feet away and could not be seen from the bridge.

The wave over wash being taken on the *Gloucester*'s port quarter would roll up onto the main deck and tend to partly flood some of the lower staterooms, pushing the ship off course and in toward land. Captain Howes would wait for the slightest lull in the waves and work the ship back offshore. Out ahead, Captain Howes had been listening for the sound of the Pollock Rip Shoals buoy, and saw it almost too late, nearly running over it.

Shortly after that, the visibility improved and the light atop of the *Pollock Rip Lightship* was sighted. This was the turning point in the voyage and he changed course about 80 degrees to starboard and began the run through the Vineyard Sound between Hyannis and Nantucket Island. First, Captain Howes would have to announce the change of course over the ship's bell and whistle system to alert the crew and passengers that the ship was about to begin a large roll to port when turning the corner south of Monomoy Island off Chatham, which was the captain's hometown.

As the ship yawed back and forth through the raging seas, the *Gloucester* picked up the red light signifying the Cross Rip Lightship moored on a station off of Hyannis on the route to Nantucket." In doing such dead reckoning, it is very easy to come in from the fog and snow, actually colliding with a lightship. In 1934, this is exactly what happened when the White Star liner

Olympic, sister ship of the ill-fated *Titanic* was coming through the fog, homing in on the Nantucket Lightships radio beacon. Appearing suddenly out of the fog, it rolled the *Nantucket Lightship* down under its keel, sinking the lightship, with the loss of seven crewmen. The practice of homing in on the ship itself instead of slightly to the side of it was officially forbidden from then on.

Next to be seen of the navigational landmarks would be West Chop Light on Martha's Vineyard. After that, Gay Head, on the western end of Martha's Vineyard, then Block Island Light, Montauk Light at the eastern end of Long Island and Fire Island Light. The visibility was getting better toward the end of Sunday night. Captain Howes headed due south following the shore and sea buoys and surprisingly arrived in Norfolk at Monday morning exactly on time. The shipping line officials were astonished. They expected that if the ship were not lost in the storm, it would arrive hours late, making the accomplishment of Captain Howes's navigation even more amazing.

Far at sea to the north, the *Portland* began to experience damage to its superstructure and deckhouses. Only two weeks' prior, the *Portland* headed out to sea in a storm, sustaining no structural damage. However, this was a storm like no other.

Not only was the *Portland* fighting for its life, but also many other ships under both sail and steam. Twenty miles off Cape Cod, the 344-ton ship *Albert Butler* under the command of Captain Frank Leland, was traveling from Jamaica to Boston with a full load of logwood. She had been enjoying a leisurely sail, scudding along on a pleasurable southeasterly light breeze.

A short time later, the wind direction changed to a northeast gale and in typical fashion, Captain Leland started sailing farther away from the Cape. He did not want to be driven ashore onto the Outer Cape beaches with their treacherous shoals.

The seas had now become so huge that he had to shorten sail until only the storm sails were flying; just giving him enough power from them to maintain position offshore. Captain Leland then positioned the *Butler's* bow into the wind and slowly drifted sternward about 15 miles.

Now, with the ship almost under bare poles, the sea at 7:00 a.m. rolled over, breaking loose the deck cargo of logs and turning them into battering rams, tearing out rigging, smashing cabins and wreaking havoc with the booms and sails. Immediately, the *Albert Butler* became completely out of control and was at the mercy of the uncontrolled ocean. Shortly after 10 am, the forward part of the keel struck bottom and pivoted the Butler around to the left and swung her broadside and threw it high up on the beach as if it were a toy. With the hull of the ship grinding slowly up the beach with each succeeding breaker, she was observed ashore by Surf man BF Henderson from the peaked Hill Bars Lifesaving Station and Benjamin Kelley from the High Head Station, who had just met at the extreme ends of their nightly beach patrol.

They agreed that this was the worst storm they had ever encountered in their lifetime. The two lifesavers immediately ran in opposite directions to alert their stations and to bring the rescue apparatus to the scene as quickly as possible. Despite the fierce stinging wind, tossing sand and ocean spray, the two men reached their respective stations 3.5 miles away in about 30 minutes.

The beach carts were instantly hitched to horses and the lifesaving teams from both stations started for the location of the wreck. Reaching the area of the disaster at about 11 am, the rescuers could see that the *Albert Butler* was still about 150 feet from the water's edge. The breathless lifesaving crews from both stations set up a Lyle gun, or small cannon, to fire a line tied to a small brass projectile and landed a line across the ship's rigging. The crew of the Butler, to the surprise of the surf men, made no attempt to make it fast to the ship. On the second try, they landed a line right on the deck of the Butler a short distance from one of the sailors, who grabbed the line and started hauling away on the whip, or messenger line, which was attached to a heavier line that would be made fast to a heavy part of the ship. Unfortunately, a couple of the sailors tried to go ashore on the smaller whip line, just as a heavy ocean wave crashed and ripped the two sailors from their tether, drowning them. Three sailors had managed to get ashore successfully through the raging surf. Later, at 2 pm, the tide and the surf had subsided somewhat.

The surf men had gotten aboard the *Butler* and found two more survivors inside the cabin, raising the number of rescued sailors to five. The last two were then carried to the Peaked Hill Bars Lifesaving station and joined their shipmates. The *Albert Butler*, sitting high on the beach, was later declared a total loss.

Farther up the coast the schooner *Mertis Perry* had smashed ashore at Brant Rock in Marshfield. Under the command of Captain Joshua Pike, she was headed back to port from the Grand Banks with a full load of thousands of pounds of fish and was very low in the water. While crossing Massachusetts Bay on Saturday night, Captain Pike chose to turn the ship and work the vessel to a more offshore course. To do this, he would have to jibe, or turn, across the wind while running downwind.

The maneuver turned into a disaster because as the forward boom swung across the wind, the gaff that holds the boom to the mast snapped off, dumping all of the wind out of that sail, then slamming and ripping the forward jib sail. This caused the vessel to come to a stop in the ocean, swinging her around into the wind and damaging rigging. Captain Pike attempted to do the only thing left, to let go both bow anchors in the attempt to stabilize things.

The two anchors held until Sunday morning, when the port anchor line broke from the strain. Luckily, the starboard anchor held fast until daylight when its line also parted. All that was left for the captain to do was to run the ship before the waves and head for shore.

With no sails or anchors left, the ship was driven high on the beach. The captain and most of the crew were able to climb down to the beach from the long overhanging bowsprit. This feat saved 10 of the 15 crewmen, the rest drowning as the ship swung sideways in the surf. The surviving crewmen were shortly rescued by the lifesavers from the Brant Rock Lifesaving Station.

CHAPTER 7

Joshua James Legendary Lifesaver

—

By now the gale was being felt from Maine to New York, for when 1,000 miles of ocean becomes stirred up and pushed west by 100-mile-per-hour winds, it swallows and destroys everything in its path.

In 1785, the all-volunteer Massachusetts Humane Society was formed to help shipwreck victims. They immediately began to set up a series of crude huts on deserted beaches that were empty except for candles, tinderboxes, firewood and non-perishable food. These huts were designed to shelter the shipwreck survivors for a few days, or at least until help arrived. The first humane society house was built on Lovell's Island across the Narrows Channel from Georges Island, the future home of Fort Warren, which would figure prominently during the future Civil War. This first humane house was erected in 1789. Eighteen years later, the first lifesaving station in the country was erected in Cohasset about eight miles away from the hut on Lovell's Island. The US Lifesaving Service specially trained boat crews that could go out in almost any weather. Inside the station was a fully equipped 30-foot double-ended surfboat that sat on its own specially designed horse-drawn wagon with steel wheels that could be driven right into the surf to float the boat.

During the following 40 years, the Massachusetts Humane Society built and staffed 18 other stations manned by crews from the U.S. Lifesaving Service. Included in this number were 12 stations along the outer Cape Cod beaches, Wood End, Race Point, Peaked Hill Bars, High Head, Highland, Pamet River, Cahoons Hollow, Nauset, Orleans, Old Harbor, Chatham and Monomoy. A later station, Monomoy Point, was added in 1901. These stations were usually about five miles apart, so that the surf patrols that ran around the clock would not have to go too far to summon additional help in a major disaster. During the *Portland* Gale of 1898, those fearless life-saving crews saved more than 133 lives under almost impossible conditions.

The Northeast gale also wreaked havoc with the various lightships stationed off the Cape that had no choice but to ride out the gale. The *Hen and Chickens Lightship*, broke free of its mooring in Buzzards Bay and was found adrift a week later 25 miles southeast of Nantucket Island.

The *Handkerchief Shoal Lightship* broke free from its huge anchor southwest of Monomoy Point and drifted five miles away. It would have drifted a hundred miles out into the North Atlantic had it not let out another anchor to drag across the bottom.

She was later towed into Hyannis Harbor for a new anchor and other superficial repairs. Southwest of Monomoy, the *Pollock Rip Lightship* broke free and after a search down the east coast, was finally located in Delaware, hundreds of miles to the Southwest. The lightship crews had the most hazardous duty of all. Each of the seven lightships off Cape Cod suffered some damage.

Far at sea, the four-masted schooner *King Phillip* was fighting its final losing battle with the elements. With each succeeding wave, more of the ship's spars and rigging would be swept away until the ship became unmanageable in the 40-foot seas. Even with small storm sails, the combination of wind, seas, and snow combined with almost zero visibility, wore ship and crew down to the point of total immobilization. The *King Phillip* was last seen off of the coast of Maine, but sank approximately 30 miles off Cape Cod. Parts of the ship washed up a week after the storm inside Cape Cod Bay. A large pump from the ship was used to identify the wreckage. After the gale subsided, wreckage from numerous wrecks continued to pile up all along the outside of Cape Cod, as well as on Nantucket and Martha's Vineyard, where many ships that had sought shelter from the storm had dragged ashore and had been smashed to pieces by the surf.

Farther up the coast in Hull, just to the south of Boston, the Point Allerton Lifesaving Station was under the direction of the world champion of all lifesavers, Captain Joshua James. Gliding past Point Allerton, the four-masted schooner *Abel Babcock*, bound from Philadelphia to Boston with a cargo of coal, attempted to anchor in the lee of Boston Light to ride out the storm. The 812-ton vessel held on for a short while, but the anchors began dragging and she was pushed obliquely across Nantasket Roads to smash to pieces on the dreaded Toddy Rocks near the lifesaving station. Toddy Rocks lie about half a mile north out in the channel from the Hull shoreline at Stony Beach. In the intensely driving snow, it was impossible for the patrolling surfmen from the station to see the wreck, which broke up immediately, killing the entire crew of eight. No one knew of the wreck until a large portion of the hull came ashore near Windmill Point at Hull Gut, which marks the western end of the long Hull peninsula.

The next morning, the three-masted schooner *Henry R. Tilton* drove up high on Stony Beach, parallel to the shore. The Point Allerton Lifesavers promptly rescued the crew. James was fearless, as were a number of his crew. At approximately the same time, two barges owned by the Consolidated Coal Company were under tow by a tugboat that could not handle them in the raging seas. The tug's' captain ordered them cut loose off Point Allerton at the entrance of Nantasket Roads. *Barge numbers one and four* were bound from Baltimore to Boston. *Barge One* became wedged on the rocks just off Windmill Point with Captain Joshua Thomas and his four crewmembers clinging to the top of the deckhouse. This soon broke free from the rest of the barge with captain and crew thrown into the surf line. Risking the frozen water temperatures, James's crewmembers tied a line around them and waded into the raging surf.

They then grabbed onto the helpless sailors and held tight while others ashore pulled in the lines to snatch the seamen from the jaws of death. Later, *Barge Four* hit Toddy Rocks and broke up immediately. The captain and one of the crew were saved after clinging to a section of the deckhouse, but the other three crewmembers were lost.

As the lifesaving station was more than a mile away, Captain James decided that immediate help was needed. Less than 100 feet away was a boarded-up summer cottage. James ordered the home broken into so that the freezing and almost dead survivors could be immediately attended to. All of the frozen clothes were stripped off the men and they were rubbed down to thaw them out, then were wrapped in blankets until they were able to move to the lifesaving station to be placed by the fire and fed warm liquids. The owner of the cottage was later compensated for the damage done by the lifesavers in the emergency. On Sunday afternoon, at about 4:00 p.m., Captain James and his crew returned to the station completely exhausted after their many rescues of the day and night before. All this time, in the face of wind, wild seas, and driving snow, things only continued to get worse. Shortly before dark, Captain James noticed the masts of a schooner very close in near Boston Light about a mile away. He expected that another ship was in trouble, but the exhausted crew would never be expected to row that distance against the fierce wind and waves. The schooner was the *Calvin Baker*, coming from Baltimore to Boston with a large cargo of coal for the region's furnaces. It had run ashore, grinding to a crashing stop only 75 yards from the lighthouse. The second mate grabbed a line and dived overboard as a measure to help the crew to safety but he was knocked unconscious and drowned in the raging surf. His body was then washed out to sea. The seven remaining crew-members climbed into the rigging to escape the billowing seas that washed completely over the decks. The next day, they climbed down out of the rigging; huddled together to share what remained of the warmth that was left, hoping for a rescue. Three hours later, the rising tide and seas forced them back into the rigging where the men were beginning to lose hope.

Early the next morning, Captain James commandeered the local tug *Ariel* to tow the lifeboat out to Boston Light. The tug stood a quarter mile off the light, while the surfboat crossed over the nearby Great Brewster Spit, a sandbar that was less than three feet of roiling water, to reach the wreck of the Calvin Baker on the other side. During the preceding night, both the steward and first mate had died, leaving only five freezing cold sailors left alive. Captain James's life-saving crew lowered them from the rigging and rowed them to the waiting tug, then headed back to the Point Allerton Station. As with the other survivors from other wrecks, these frozen men were also stripped of their icy, board-stiff clothes and wrapped in warm blankets. They were slowly fed hot soup to help thaw them out. After about four days of this care, they were free to journey home to their loved ones.

Across Massachusetts Bay, 27 schooners had made it into Provincetown Harbor to ride out the gale. At the height of the storm, wind gusts had reached more than 100 miles per hour. Waves inside the harbor continued to grow higher and steeper causing the ships that had sought safety to strain mightily at their huge rope anchor lines.

At the height of the storm, the wind turned from northeast to northwest, exposing the outer part of Provincetown Harbor to the full winds, but not the highs seas because of Long Point, which extends all the way around to form a large hook heading west, then going around to a southeast direction. Even then, the northwest winds caused 10 of the larger schooners to part their anchor cables, which were made of heavy manila hawser, to pound ashore, or to sink at their anchors.

The smaller vessels drifted into shallow water and, had their bottoms pounded out. Ashore, nine huge wharves collapsed, while 21 buildings were blown over. The scene of destruction in Provincetown was almost beyond belief. In Boston, the driving snow began to pile up into three-foot high drifts, as the hurricane winds drove the snow everywhere. The huge storm surge washed out the foundations for train tracks here and there, while knocking over telegraph poles and felled huge trees across the tracks. Hundreds of private vessels were ripped from their moorings and thrown ashore in heaps, many smashed beyond recognition. In a storm of this magnitude, visibility is barely 50 to 100 yards.

The more than 300-foot Wilson Line steamship *Ohio* entered the harbor almost blindly, went out of the main channel and drove high up on Spectacle Island. The pilot, Captain William Abbott, said that in the white-out conditions he could barely see across the ship. It took a week for the tugs to finally pull her off the beach and into a nearby dry-dock to repair the damaged bottom of the steel liner. In Gloucester Harbor, as many as 30 vessels were either wrenched from, or sank at, their moorings. On Boston's South Shore, the schooner *Frederick Alton* smashed ashore after dragging her anchor and shortly became a total loss. At George's Island in the outer harbor, the two-masted schooner *Lizzie Dyas,* loaded down with granite from Maine, struck the rocks off the eastern shore and ground herself to pieces, also a total loss. Even if a ship went ashore on soft mud, the huge, billowing waves would shortly finish it off, crushing in the sides, while washing the crew overboard. This was mostly fatal due to hypothermia, as it is known today, plus, few could swim. Today, professional training for situations such as this is widely available.

CHAPTER 8

Struggling to Survive

FOR THE *PORTLAND*, THE STORM'S intensity increased after she had passed Thatcher's Island Light to port and had proceeded to the northward toward the city of Portland. The snow thickened quickly. Not to be caught on a lee shore, the *Portland* began to work its way out to sea to clear the coast and find some maneuvering room. Making headway against this bad gale was near to useless, for all that would have been gained would be to spring more leaks and use more of the precious reserve of coal. All one could do in such an intense situation was to keep the bow, or front of the ship, into the sea to lessen the damage and to keep passengers as comfortable as possible. In seas this huge, the wooden hull of the ship would twist and groan as heavy timbers in the frame would rub slightly against each other.

At nightfall, the driving snow grew heavier all the time. The addition of the sheets of freezing spray caused the steamer to become top-heavier, with the weight of ice now sticking to the ship's superstructure. This alarming weight increase caused the ship to roll more and more, along with the problem of windows of the wheelhouse icing up.

This ice had to be continually scraped away by the crew for any forward visibility to be maintained at all. The rising waves were, by now, coming completely over the 35-foot high bow of the ship and were smashing against the lower part of the forward superstructure, causing minor damage. This in turn, started to loosen the joint where the superstructure attached to the main deck, starting numerous small leaks into the ship. This was usually taken care of by the ship's main bilge pumps, but as the leaks became worse, the pumps had to work much harder, taking more power away from the main engine. This, in turn, called for more pressure from the boilers, resulting in more coal being used. The bags that were stacked on the deck are the first to be added to the main bunkers to keep the level up and not to be lost overboard. The stokers and firemen now had to shovel the coal into the boilers at a faster rate, exhausting the stokers, so much that other crew members were brought in to assist.

It now took the combined strength of three men to wrestle with the dual six-foot diameter main steering wheels just to keep the rapidly tossing ship on course, gradually working eastward offshore. In its original design, working schooner rigged masts, fore and aft, were installed on the *Portland* in case of an accident to the main machinery. However, trying to set them in these conditions was difficult, and they could only be used to steady the ship. Even though Captain Blanchard had been promoted to the rank of captain only two weeks before, he had sailed many trips on the *Portland* as first pilot, knowing how the ship acted in a gale. Now all of those years of knowledge would be brought to bear on the situation.

In an extreme emergency, there was one last desperate measure that could be tried to get even more than 2,000 horsepower from the engine. A wooden wedge could be inserted above the cutoff valve to keep it shut and build pressure above 80 pounds. In the eyes of the Steamboat Inspection Service, this is clearly illegal and could lead to a catastrophic boiler explosion, killing many passengers and crew, but it shows what could be done in a truly desperate situation. By now, the wave heights reached 40 feet and the winds over 90 mph, no wooden ship built was ever made to withstand this kind of punishment. If the ship could be kept bow to wind for the remainder of the night, during the next day, the ship might do some backstepping, or floating downwind. With an increase in visibility, the ship could be worked to the westward slightly and run into Gloucester Harbor.

All efforts now focused on not taking too much damage until daybreak, which should occur shortly after 5:30 a.m. battling huge seas fourteen miles offshore in a gale of driving snow and wind gusts of more than 100 miles per hour would wear out any captain and crew in a matter of hours. But in this extreme emergency, a person has to call upon the strengths of pure adrenaline just to keep going. In recorded history, many people of ordinary strength have given superhuman effort. In the Olympics, this one fact distinguishes those who earn the gold medals. The huge eye of the cyclone was heading right toward the tip of Cape Cod at Provincetown, the U.S. Lifesaving Service Stations on Cape Cod and elsewhere were out in the worst conditions of their lives. Normally, walking their lonely beach patrols was desolate enough; however, the conditions they endured were brutal. Without face-masks or goggles, just walking while looking out to sea with 90-mile-per-hour winds, along with driving snow and sand pelting their eyes, made their job extremely difficult.

All during the night of Nov. 26 into the 27, only flotsam, or junk, that would have floated off a beach at an extremely high tide or storm was seen on the outer Cape Cod beaches. The next night, this would all change dramatically.

On the *Portland*, conditions were holding steady. The ship was being kept head to sea, the damage was not getting any worse and the coal supply was holding out, mainly due to the extra supply that had been put aboard before leaving Maine. Having been used for almost 90 years, the paddlewheel type of steamer was gradually nearing its end. Having been designed mainly for use in rivers and harbors, in any kind of rough sea condition, it was at a decided disadvantage.

According to W.H. Bunting, in his book, "Boston: Portrait of a Port," "the disadvantages were serious. Wind damage was great. In cross seas, some of the steamers were notorious rollers, and no doubt many seasick passengers, watching the arc described by the clothes on the coat hook, became convinced that the only force preventing the steamers from going completely over was the fragile buoyancy of the overhanging 'sponsons.' "It is said that in rough weather it was often necessary to shift cargo to the leeward in order to elevate the windward sponsons away from the smashing seas, although this procedure rendered the windward paddlewheel nearly useless. In severe weather, it was often impossible to keep a steamer head to wind. One candid Boston-Bangor skipper admitted. She is only fit for smooth water, and my greatest care is not to get caught out with her on the first part of this route. If I do, it means getting out of the wet at the first chance and favoring her in every way, shape, or manner. Once let a sea strike under those infernal sponsons and it would start the whole top hamper- superstructure," he said.

This would have indeed happened by now to the *Portland* during its brave fight for survival throughout the night, but the pumps were able to just keep up with it thanks to the superhuman efforts of the crew. At about 6 a.m. the first rays of light began to light the eastern horizon, showing the full terror and majesty of the storm. If the storm looked hairy at night, it became terrifying by day, for now all the waves could be seen hitting the ship as well as the ones behind them. There was some ray of hope though, due to the possibility of a government rescue ship would see them and possibly stand by to render assistance. Plus, some light would at least allow the crew to see better and start repairing some of the damage to the ship's large superstructure, as well as plug some of the numerous leaks between the deck and deckhouses.

Also with better visibility, the bow of the ship could now be turned slightly toward land and hopefully sidestep the huge seas, by slipping into a temporary safe harbor anywhere along the coast. This would be the foremost thought on the ship's officer's minds. All this time, a double clapper engine room gong would relay the commands from the wheelhouse to the engine room. This would be used to change the engine speed or alter the direction in case of having to back down in an emergency, such as a near collision. In 40-foot seas, however, backing down was not an option, for that would allow the ship to fall off the waves and capsize. The best thing in this situation would be to try to avoid it, or at worst take a glancing blow, to keep the damage to a minimum.

Early that morning, three ships were off of the tip of the Cape, fighting the huge seas. The largest and most seaworthy of the trio being the new 297-foot *Horatio Hall,* under the able leadership of Captain Albert Bragg of the Maine Steamship Company. The *Horatio Hall* sailed from Portland the evening before at 8 p.m. and was steaming toward its destination of New York, but had to heave-to into the wind, just marking time at midnight when the wind and seas were expected to subside somewhat. Being constructed of steel and fully loaded, made her ride the seas in an easier fashion than a smaller, lighter vessel.

The second and next largest was the wooden screw steamer *Pentagoet,* which at 128 feet was operating as a freight steamer between New York to Rockland, Maine. The third ship was an 89-foot, two-masted schooner called the *Ruth M. Martin,* under the command of Captain Michael Hogan, who sighted what he thought was a large "side-wheel steamer" away to the southeast, but in actuality, the *Portland* as still 24 miles to the north of Cape Cod in her final throes of her battle with the sea.

What Captain Hogan actually saw was the black hull of the *Horatio Hall,* completely covered with salt and ice with a large wave washout down the hull, which would appear black and strongly resemble the paddle box of a side-wheel steamer. The most accurate way to look would to be to count the number of smokestacks to be sure, for the *Portland* had two side-by-side stacks, while the *Horatio Hall* only had one. Nowhere in his reports of the storm is the number of smokestacks noted. He later testified to this at the Portland storm inquiry held in 1899. Captain Bragg later testified that the *Horatio Hall* was indeed off Provincetown that morning and did not blow off four short blasts of his steam whistle, signifying an emergency. At 5:45 a.m., Captain Samuel O. Fisher of the Race Point Lifesaving Station heard four short blasts from a steamer's whistle-. It must surely have come from the steamer *Pentagoet.*

The raging seas would have by now smashed all of the wheelhouse windows of the *Portland*. The combination of the wind, snow, huge seas, and ice were making the *Portland's* deck surfaces frozen and slick. If that were not enough, the main steering wheel would become slick as well. Just keeping the ship's bow into the wind was all that the efforts of three men could do.

The reason that Captain Hogan was able to even see any ships at all was due to the fact that at approximately 9 a.m., the eye of the storm passed over the tip of Cape Cod, causing the churning ocean to moderate slightly, the sun to some out briefly, and visibility to improve dramatically.

CHAPTER 9

The Dark Clouds Gather

—

ALL THROUGHOUT THE TERRIBLE NIGHT, the crew of the *Portland* labored mightily to keep her bow into the pounding seas, with the twisting and groaning that a great wooden vessel makes, the inevitable leaks start to get worse and worse. By now, so much steam pressure had to be diverted from the engine to run the pumps, that if the ship is not to make port immediately, the ship would be swallowed by the sea. Near 9 a.m. on the morning of Nov. 27, the waves began to break away the portside paddle-wheel guards. Even though there are no surviving witnesses to the *Portland's* sinking, this following account, using some knowledge of the wreck's position and condition, surmises what happened.

After valiantly fighting the near impossible seas for more than 11 hours, Captain Hollis Blanchard and his crew were nearing total exhaustion. One can only give one hundred percent effort for so long before the mind begins to break down. With the entire pilot house windows smashed out, one smokestack leaning over slightly, the superstructure heavily damaged, death in the freezing sea was now a foregone conclusion. The writer suggests a scenario which might have happened to the ship

Ominous noises began to come from below between the paddlewheels. After running at above maximum effort, the main paddle-crank-rod bearing began to pound and shake, causing much vibration from the paddle shafts. A two-piece bearing wraps around the crankshaft-throw to transfer the engine horsepower to the main paddle shaft. As the pounding got louder, an engineer tried to pour heavy oil on the bearing journal, but to no avail. One of the bolts holding the bearing halves together split and the eight-inch diameter crank rod ripped loose from the paddle shaft, causing it to flail about the engine room, destroying everything in its wildly swinging arc. This immediately caused the engine to run away, or over-rev, damaging the main cylinder.

As the paddlewheels slowed to a stop, the ship turned to starboard and went broadside to the huge seas, which immediately smash the portside of the lower superstructure. Passengers were hurled from their berths and crushed by the falling timbers from the frame of the superstructure. The next sea then ripped the already weakened forward superstructure off the ship, dumping many passengers and crew into the freezing ocean. The superstructure stayed in mostly a large forward section and hung motionless just underwater. The main hull, with all of the heavy machinery holding it down, began its downward journey to the bottom, hundreds of feet below.

As the cold Atlantic Ocean hit the red-hot boilers, one seam of the boiler buckles, allowing for a huge amount of steam to escape, just as the main hull went under water.

The ship, still turning to the right, sank in a corkscrewing motion to the bottom of Massachusetts Bay in only minutes. On its final approach to the bottom, it leveled off underwater and floated down gently in a rocking motion, like a leaf falling from a tree. It then came to rest on the soft muddy bottom more than 400 feet below. Humans would not see it again for 91 years. The advent of new technology made it all possible.

The few people who survived the sinking were now floating on the water, battered by waves. Most of the remaining crewmembers who survived the escaping steam now were trapped in the main part of the wreck, more than 400 feet down. The passengers left afloat in the 30-degree water now began to lose consciousness, one after the other.

The intense cold went to their limbs, head, and then torso. One by one they died, never to come ashore, for the vengeful sea has now taken everything. The superstructure, which was still barely afloat, began its quickening drift with the current (not the waves) south and slightly out to sea from the sinking hull of the ship. Over a period of 13 hours it drifted south by east, a distance of 24+ miles on the outside of Stellwagen Bank dropping off pieces as it went; The superstructure hung below the water to a depth of 50 feet while the pilothouse end was on or just below the water's surface. South of Stellwagen Bank, it drifted slightly westward then eastward again as the wreckage neared the dreaded Peaked Hill Bar reef just off the northern tip of Cape Cod. Near 10 at night, the fifty-foot-deep part of the superstructure started bouncing off the thirty-foot-deep Peaked Hill Bar, forcing the upper section up out of the water into the full fury of the waves. The superstructure now began to break up, tossing cargo, milk cans, and bodies out of the wreckage. A short time later, the superstructure broke up completely and the waves tossed it up on a half-mile long section of the beach near High Head Station. More of the wreckage, including bodies, ship's wheels and the: pilot house, are carried by the eastward current close to the beach east and to the south as far as Nauset and Monomoy Island.

The wreckage from two other ships, the *Pentagoet* and the two-masted schooner *Addie Snow*, also are thrown ashore, mixing with the *Portland's* wreckage. No bodies from either of these ships ever washed ashore. The sea guards its secrets only too well.

As early as Sunday morning, officials in Portland noted that the steamer had failed to dock at its usual time. However, due to the terrible gale, she would be expected to be late. After 6 hours went by with no sign of the ship, worried officials of the Portland Steamship Company began to call the cities of Boston and Gloucester to check if the ship had docked there.

The intensity of the storm knocked down many telegraph poles along the coast, so no news could be obtained from Cape Cod or the islands. There was hope in Boston and Portland that the *Portland* might have taken shelter in

Provincetown Harbor, but that news would have to wait until train tracks and roads were cleared. Liscomb, the general manager of the line, called up and down the coast all day on Sunday, the 27th until 10pm, and then gave up. The next day, the 28th, he contacted the U.S. Revenue Cutter Service and arranged to hire the cutters *Dallas* from Boston and the *Woodbury* from Portland to conduct a search as soon as the seas subsided.

Late Sunday night, and well into Monday at the offices of the Portland Steamship Company in both Boston and Portland, the relatives of the passengers and crew of the *Portland* gathered to find any information. The officials at both locations knew nothing more than the relatives. The search by the *Woodbury* found no trace of the ship, but the *Dallas* had better luck, finding some of the wreckage floating off the Peaked Hill Bar at the tip of Cape Cod.

"On Monday, Charles F. Ward, the Chatham Correspondent for the Boston Herald was in Hyannis when a telegraph message from an assistant in Truro started to come through. The wire went dead, but not until the fact that the *Portland* had been wrecked was communicated," according to Thomas Harrison Eames in "Steamboat Lore of the Penobscot." Connections all along the Cape were broken, but realizing the importance of the news, Ward boarded a train that left Hyannis at 6:30 p.m. At East Sandwich, the train was stopped by a washout and he struck out for Sandwich on foot through the snowdrifts. Arriving at 11:00, he hired a horse and rode to Buzzards Bay, where he was able to catch an early train for Boston with the first news of the disaster.

Tuesday forenoon, a message from the Herald office to Agent Charles Williams of the Portland Steamship Co. in Boston ended the long suspense of the Company officials and those who were waiting in the lobby.

At 4:00 p.m. that afternoon the *Longfellow* arrived from Provincetown with more information about the tragedy. One of the passengers, E.H. Cook, asserted that on the previous morning he had seen the "top" of the *Portland* thrown high on the beach near Peaked Hill Bar Lifesaving Station. This was actually High Head in Truro. The Boston Herald chartered a Commercial Wharf tug and sent a number of reporters to Provincetown to gather details on the wreck. With them went Agent Williams of the Portland Steamship Co. and William Peak of Barnstable, whose employer, Charles Hersom, had commissioned him to look for the bodies of his son and daughter-in-law, who were passengers on the *Portland*.

The sea was smooth and oily, but with a huge swell. At the two funeral parlors in Provincetown, Williams was able to identify some the ship's personnel. At Orleans, also, bodies were coming ashore and Mayo's, a blacksmith shop, was turned into a temporary morgue. Dr. Samuel T. Davis of Orleans supervised the identification of the bodies washed ashore, aided by photographs that Williams requested from friends and relatives for that purpose. The railroad station at Orleans was piled with coffins in which identified bodies were shipped to relatives. Others were sent to the North Grove Street Mortuary in Boston.

The most notable piece of the ship's wreckage was a six-foot high double wheel that was lashed together had come ashore at Orleans. Eames, in "Steamboat Lore of the Penobscot," believes that "the fact that it was lashed gave rise to rumors that the officers had lashed the wheel when the *Portland* could be managed no longer, but Captain Benjamin C Sparrow of the Cape Cod Lifesaving Station exploded these theories by pointing out that the lashings showed that it was the spare wheel, which would have been used only if the main steering apparatus had broken down."

Since that time, numerous pieces of the ship have been recovered by fishermen or have washed ashore. In an odd coincidence, in April 1899, the schooner Maud S., one of the vessels that passed the *Portland* on her final trip, was trawling on Stellwagen Bank and brought up some stateroom fittings and electrical equipment known to have been part of the steamer.

A trawler recovered one of the *Portland's* brass lanterns twelve miles southeast of Thatcher's Island, near the point that Captain Pellier sighted her during the storm. The lantern probably was one of the first articles torn from the steamer by the raging sea. Another trawler found one of the steam gauges on Stellwagen Bank, and a chandelier from the vessel was nearby.

In June 1924, Rockland, Maine, Captain Charles Carver of the scallop dragger *Harriet Crie* was dragging the bottom approximately nine miles to the north west of Highland Light. In his trawls were brought up a number of pieces of wreckage, including broken dishes, bent silverware, a stateroom door lock, and brass fittings. These items, almost a century later would be called a debris trail which could be followed by modern electronic means to find the main body of the wreck.

In all, only 40 bodies were ever found of the 192 souls lost aboard the *Portland*. Most were found immediately after the storm, but some, more heavily sanded in bodies were unearthed on the outer Cape Cod beaches during another storm in early December. In the ensuing weeks after the ship was lost, many of the newspapers on the Eastern seaboard heavily criticized the Portland Steamship Company for not conducting a search for the ship. Although it had chartered two cutters to search from both Boston and Maine the steamship line did not publicize their efforts.

CHAPTER 10

The Storm of the Century

BY FAILING TO MOUNT ANOTHER immediate search for the *Portland* farther out to sea, the management of the Portland Steamship Co. raised the ire of the local and some of the national newspapers.

Even if floating wreckage had been located, any potential survivors likely would have been dead from hypothermia. In the first half of December, The Boston Globe newspaper could no longer wait for more concrete news and organized and paid for a week-long search at sea for more of the *Portland's* wreckage.

Navy Lt. Nicholas Halpine was in charge of the U. S Hydrographic Service office in Boston. He was most familiar with the ocean currents in Massachusetts Bay and knew the sea-bottom characteristics better than anyone in the area. The Globe arranged for the charter of two Boston tugboats, the *Gallison* and the *Chesterton*, and equipped them with drags and divers in case the wreck was located on the bottom inside the Peaked Hill Bars reef to the north of Cape Cod. Due to high sea conditions, the newly arrived tugs were not able to begin dragging a cable between them off the Peaked Hill Bar until Monday, Dec. 13.

A half-mile length of heavy chain strung between the tugs a quarter of a mile apart would be dragged along the bottom. If the chain snagged, divers in old-fashioned hard hat equipment would be sent down the chain to identify the possible wreck. Visibility at the bottom is often only 18 inches. In the case of the *Portland*, its sinking was so recent that conditions for recovery of gear and bodies would be most favorable.

Dr. Joshua Lewis of the local board of charity, indignant over the failure of the steamship company to look for victims farther out to sea, met with Massachusetts Governor Roger Walcott and asked for his help in starting a search seaward to the east and south of the Cape, including Nantucket and Martha's Vineyard. Shortly after this, the tugboat *Herald* left Boston to begin the search. Three police detectives and two newspaper reporters joined the tug's crew for the search. Over a three-day period, the tug cruised the entire outer Cape, all the outer shoals, Vineyard Sound, and back, covering a distance of 392 miles, and no bodies or even a scrap of wreckage were sighted.

As most of the watches found on the victims had stopped between nine and ten o'clock, 9:00 p.m. was thought to be the time of the sinking, suggesting that the wreck location would be off the Peaked Hill Bar, north of the Cape.

Since Lt. Halpine's search failed to find the wreck on or inside the Peaked Hill Bar, the wreck was believed to lie in deep water somewhere to the north. In 1899, the fishing schooner Maud S. suggested a wreck site 20-plus miles to the north in Massachusetts Bay.

One other casualty of the storm was the telegraph cable that ran from Duxbury, Mass. to the Maritime Provinces of Canada that included Nova Scotia, Newfoundland and the island of St. Pierre. The cable-laying ship, *Minea*, was given the job of repairing the broken cable, which had parted at a point some seven miles to the northeast of the dreaded Peaked Hill Bar. It was hoped that the wreck of the *Portland* might be found during the repair effort, but there was no sign of the wreckage when both broken ends of the cable were brought up from the bottom to be spliced back together.

As maritime historian Edward Rowe Snow later suggested in his extensive research on the disaster, "If the wreck had sunk that far to the north of the Cape, the waves would have scattered the wreckage all over the southwestern beaches of Massachusetts Bay, not on a half mile stretch of beach near High Head in Truro where the majority of the wreckage was concentrated."

Later, during the latter part of World War II, after Mr. Snow was convalescing from being wounded in action in North Africa, he journeyed to Rockland, Maine, to interview Captain Charles Carver about the exact location of his find of *Portland* artifacts. Capt. Carver gave Mr. Snow the exact navigational coordinates to the spot where he dragged up the artifacts, after which Mr. Snow started to save money for a later diving expedition to the site. He also came to the conclusion that the *Portland* might have collided with the heavily laden Rockland-based schooner *Addie E. Snow*, the wreckage of which was mixed with the *Portland's*.

After raising the considerable sum of $1,800, Mr. Snow sought out local diver Al George, and the salvage ship, *Regavlas*, was soon chartered for this purpose. Surveys were made during June and July 1945.

"In the month of June 1945, I was commissioned by Lt. Edward Rowe Snow to descend to the bottom of the ocean off Cape Cod at a location previously found by Captain Charles G. Carver of Rockland, Maine, George reported on the expedition. Highland Light is at a distance of four and one half miles; the Pilgrim Monument has a bearing of 210 degrees. Race Point Coast Guard Station is seven miles distant.

Arriving on the location the last week of June, I carried out plans for finding the *Portland*. I ran a course 115 degrees true from the Peaked Hill Bar Buoy. I made a sweep after reaching a point one and ¾ miles from the buoy using a span of 600 feet of cable. We swept the entire location within a radius of three quarters of a mile.

On the second time across, I made fast to what I knew was some large submerged object. After buoying it, we swept the entire vicinity to make sure the object was the steamer *Portland* and not some other wreck. Of this, I am certain; this wreck is the only wreck in this vicinity, which corresponded to the bearings given by Captain Carver. Therefore, it must be the steamer *Portland*.

Realizing this fact, I got rigged for diving. I slid down the sweep wire and within three minutes of the time I had left the *Regavlas*, I had landed on the *Portland*, which was over on its beam-ends and heavily sanded in. It may surprise the average person to realize that the visibility here is less than 18 inches.

It was a weird sight. Crawling along the sloping hull of the vessel, I nosed my helmet forward until I ran into a mast heavily encrusted with marine growth (mussels, seaweed, etc.). Reaching my hands out, I found I could not span the mast. I followed the mast up until it was out of my reach, at a space between two gigantic boulders on the bottom. The mast appeared to be broken off about 15 feet up. It would seem as the *Portland* had hit bottom on her beam-ends, and then, through the years had worked its way into the sand until it was buried almost completely. Only the bare hull of the ship seems in position. All superstructures evidently have been spread around the ocean bed long ago. The boulders are much higher than my head. I could not tell whether it was the foremast or the mainmast. Going down on my hands and knees, I would make out the ripples of sand on the bottom of the sea and could see little shells from time to time. The tide was running about one knot and it was slack water. My brother telephoned down from upstairs that he had 300 feet of line run out to enable me to stand on bottom in 144 feet of water. It was a strange experience standing there alone with the ill-fated *Portland* and probably what remained of the passengers and crew still imprisoned in her sand-covered hull.

I wish I could give one the awesome picture. While visibility was a foot and a half, vague shadows could be made out up to five and eight feet away. Giant devil weed and long streamers of the other varieties of seaweed shrouded me in a big black cloud of marine life.

As there will probably be many who might think that the *Portland* sank gently to the bottom to remain practically intact for the 47 years since the disaster, I must impress on their minds the true picture of the present conditions. The entire hull of the vessel which protrudes above the sands is a blacked shapeless mass of water soaked wood, seaweed, mussels, scallops, and scores of different types of marine growth. I spent less than half an hour on the bottom, then I gave the signal to be hoisted to twenty feet from the surface where I hung for ten minutes, then I was hoisted to ten feet from the surface, where I remained for fifteen minutes. I was then brought over the side and my dive had been completed.

I realize that the purser's bell, the keys and doorknobs and the many other articles which have been brought to the surface from this shipwreck indicates many more articles could be retrieved. I have been told that a small fortune in uncut gems in the purser's safe would well repay the lucky finder. In my opinion, however, although I would be happy to undertake the search, the chances are greatly against anything more of practical value ever being found. If anyone would consider financing such an enterprise, the cost would be prohibitive."As far as Mr. Snow was concerned, this diving expedition had finally answered the question of the wreck's location. He was 95-percent sure that this was the ship, but there would always be some lingering doubts.

On hearing and reading about the *Portland* years later, many of his readers and listeners urged him to undertake further expeditions to the site. But no one ever offered to help him pay for an expedition. One can only go so far on speculation.

On a visit to Mr. Snow's office at The Patriot Ledger in 1967, I brought along a government chart of the Massachusetts Bay and asked him to draw the exact coordinates of the wreck. I promised Mr. Snow that if he could not find the rest of the material to complete his then uncompleted book on the loss of the steamer *Portland*, then I would complete it myself when the material would be found in the future using new technology that still was years away from being invented. It would take 28 more years until this endeavor was completed.

During the 1970s, the Historical Maritime Group of New England from Bourne, Mass., comprised of three superb shipwreck hunters from American Underwater Search and Survey, John Fish, Herbert McElroy, and Arnold Carr began their own search for the *Portland's* remains.

Something was glaringly missing from diver Al George's description of the shipwreck. If the wreck were heavily sanded in and on her side, and the paddle boxes had been stripped away, at least the paddle shaft should have been protruding. The *Portland's* engine had a 50-foot-high A-frame that encompassed the entire engine, atop of which should have been the massive walking beam used to transfer the up-and-down motion of the piston to the massive crank rod, which turned the paddlewheels. None of this was found in the George's dive. When underwater cameras were later lowered to the site and lit up the wreck, it was shown to be an unknown wreck of a five-masted schooner. That is why George's arms could not reach all the way around the broken-off mast. The masts on the *Portland* were not as thick as one's on a schooner, as they did not have to provide as much support as those on a schooner.

To be a success, one has to know when it is time to seek out the best help. After ruling out the Snow site and some others near the tip of the Cape, HMG arranged for world famous Woods Hole Oceanographic Institute's help in the search.

One of the oceanographers at the Institute, Richard Limeburner, took all of the weather and current data known from the storm, including the 1898 U.S. Weather Service data and fed it into the institute's bank of computers. The data suggested a search area far to the north of the Cape, east/southeast of Cape Ann, where the ship was last sighted on the night of November 26, 1898. A wide area had to be covered by both magnetometers and side-scan sonar.

Of the 27 wrecks that were scanned with the magnetometer, the best 10 would be scanned with the side-scan sonar. Underwater cameras and a core sampler would directly access the best of these. Late in 1989, the wreck was found in deep water off Cape Ann.

HMG said in a press conference that the wreck location was more than 300 feet deep, but that the exact location was kept secret.

In November 1998, a special edition of WCVB-TV's news magazine Chronicle featured interviews with Fish and Carr, who showed both still photos and video of the wreck. As the video revealed intact a steel paddle box, and the *Portland* was the only side-wheeler known to have sunk in the area, they were 95-percent sure they had found the *Portland*, more than 22 miles north of Cape Cod.

In future expeditions to the wreck, they needed to find the ship's nameplate to be 100-percent certain. After watching the videotape seven times, I called John Fish and told him that he let some information slip that showed that the wreck was deeper than originally thought. Markings on the video indicated that the top of the paddle box was at a depth of over 400 feet, and the boxes were 30 feet off the bottom, putting the true depth of the wreck at more than 400 feet.

In that conversation, I offered to make him a copy of a photo of a Portland Engine Company control panel that showed the location of the engine's #57.

The team had been looking for a nameplate with underwater cameras, but did not know the exact location.

In my research, I had asked the men in the boiler shop at Bath Iron Works to find out where the company's nameplates were located on their 1800s-era round boilers. After 10 months, they said to look to the side of the coal loading doors in the seven-o'clock position.

In addition, the ship's bell, located on the port side of the walking beam frame on the top deck would have the *Portland's* engraved on it. In a future trip to the wreck location, Fish now has three definitive targets for a nameplate.

The mystery of the *Portland's* final location, and the excitement of the search for the wreck, has inspired more than just this book.

John Fish is working on a book about the loss of the *Portland* from an oceanographic standpoint, and in addition to Mr. Snow's writings on the *Portland*, Mason Smith of Cape Elizabeth, Maine, and co-author Peter Batchelder, wrote the fine volume, "Fort Short Blasts," which came out just before the storm's 100th anniversary in 1998. It was beautifully researched and written (they both had also known Mr. Snow) and was featured in the Chronicle episode that included Fish's video.

When the dive teams confirm the *Portland's* identity by discovering its nameplate, I will be thrilled to have had a small part in helping locate the proof. Fair seas and best of luck to HMG in its future expeditions; only then will the vengeful sea give up some of the last secrets of the *Portland* saga.

CHAPTER 11

The Portland Inquiry

AFTER EVERY MARINE DISASTER COMES the recriminations for who was at fault. During the storm of the century, people and lawyers will say that warnings were ignored and that the steamship company would be solely at fault. Hardly ever would the sometimes-confusing reports by the weather service ever be brought into the discussions.

In early December, U.S. Rep. John Fitzgerald petitioned the Secretary of the United States Treasury to investigate the loss of the *Portland*. In the absence of any survivors to interview, the U.S. Steamboat Inspection service declined to investigate. Civil suits would have to settle the matter.

The first suit against the Portland Steamship Co. was filed on Dec. 23, 1898 by the father of passenger Nathan Cohen. The suit charged that the steamship company with negligence. Benjamin Thompson, an attorney representing the Portland Steamship Company, went before the U.S. District Court in Portland and asked the court that the company be ruled free of liability, as the storm was an "act of God". A petition was also filed on the company's behalf to limit damages to the value of the ship and its cargo, in the event the vessel was recovered. Judge Nathan Webb refused this last request so as not to deny future claimants. The court also advertised that suits against the Portland Steamship Company must be filed by March 30, 1899.

At the close of business on March 30, 55 lawsuits against the Portland Steamship Company had been filed, seeking more than $494,000. After this time, some further searching for the wreck was done, but nothing of note was found other than dishes, some furniture, and silverware. During the proceedings, many witnesses were called by the Portland Company, including the managing director of the New England Company that had built the ship. Other witnesses were former captains and crewmembers of the *Portland*, as well as the captains of those ships that had last seen the *Portland* on the night of the storm. Two of the most listened-to witnesses were U.S. Steamboat Inspection Service Captains Merritt and Pollster, who had only recently inspected the ship during its last dry-docking. They reported that some extra bracing had been installed and test borings concluded that there was no rot in the planks, and that she was stronger than when newly built.

Amos Haggett, superintendent of the New England Company, testified that he oversaw every step of the ship's construction and said that no financial limitations had been put on the company as to the quality of the wood or the expense of her fittings. Also called to testify was George Morse of the Portland Engine Company, who said the work had been completely and properly done and that no trouble had ever been reported about Engine #57 in the more than nine years it was in use. It was also the most powerful in terms of horsepower that the company had ever built. It could not be said that the ship was underpowered. This allowed the *Portland* the ability to steam into fierce winds when other steamers were virtually stopped in place. The Portland Company's general manager, Liscomb, testified that at the *Portland's* annual overhaul – which lasted from six to eight weeks every spring – no financial limitations were put on the overhaul crew. The ship was given whatever it needed no matter the cost.

The Boston newspapers had written about the inferior Tule life preservers that were supposed to be aboard the ship, describing them as no better that dried grass. Liscomb refuted these reports, testifying that the ship carried 758 of the much better full-length cork-filled life jackets. Each passenger and crewmember would have had three each had they wanted them. The Boston and national news accounts also had missed that the Tule life preservers existed only in belt style, and there were none of these on the *Portland*.

At the end of two days of testimony, Judge Webb issued a lengthy decision. The Portland Steamship Company was free from liability, as the loss of the ship was ruled an act of God. The company was ordered to pay court costs, and with that, the case against the Portland Steamship Company was concluded.

In the aftermath of the Portland- trial, the lost crewmembers' survivors were given a month's pay by the Portland Steamship Company to ease their suffering. Since 1844 and up until the steamer *Portland's* loss, the Portland Company had never lost a passenger in its many years of steaming the most dangerous waters anywhere.

The loss of the Portland and the unsuitability of the paddlewheels in heavy seas caused much consternation in Maine steamboat circles.

In the future, longer, higher-sided and more enclosed steel propeller-driven ships would be built in Delaware to replace the side-wheelers. These older ships would be relegated to harbors and rivers, mostly protected waters. In 1899 the *Governor Dingley* replaced the *Portland's* spare side-wheeler, *Tremont*, which was then sold to the Joy Line out of New York. Later, two more steel screw steamers, the *Calvin Austin* and *Governor Cobb*, were entered into service.

In 1901, the Portland Steamship Company was taken over by the Eastern Steamship Company in a consolidation by most of the Maine Steamship lines by C.W. Morse. In 1910, the *Portland's* near-sister ship, *Bay state*, was rebuilt. The huge radial paddlewheels and their paddle boxes were removed and replaced by the smaller, more efficient feathering paddle-wheels that took up less space, leaving deck area that could be turned into more profitable cabin space. Two more decks were added to the topside of the ship. In the future, a new regulation would be adopted, requiring that all passenger ships leave a duplicate list of all passengers ashore. Never again would the exact number of people be left unknown (except for the occasional stowaway).

In addition to carrying the latest and safest life jackets, the *Portland* also had carried eight metal lifeboats of the latest design. However, in 40-foot seas, trying to launch them would be next to impossible. Even so, there was a shortage of officers aboard. Three had been excused from making the return trip to attend the funeral of retired Captain Deering in Boston. In a later report one of the retired captains of the line, Captain Craig, said it was illegal for the *Portland* to sail short of officers. Everyone placed the blame on Captain Blanchard for the loss, but years later, in talking to many retired sea captains in Boston, Mr. Snow said the captains told him that they also would have sailed given the same information Blanchard faced at the outset of the voyage.

In many museums around Massachusetts, artifacts that were brought up or floated ashore from the ship are on display and can be seen by the general public. The Pilgrim Monument and Provincetown Museum has by far the largest collection of artifacts from the ship. The Truro Historical Society also has some of the furniture and glass decanters that washed ashore.

The Peabody Essex Museum in Salem also has artifacts donated by the Snow family. Some items are in private collections on Cape Cod, having been passed down through the generations. In 1945, Mr Snow purchased two of the doors that had washed up from the ship's superstructure, had them cut into three-quarter-inch squares and glued these inside the covers of the books he autographed. The collection of the Weymouth Historical Society Museum at the Tufts Main Library includes a lifejacket and a light bulb from the ship. The Maine Maritime Museum in Bath displays an oar and an engine room gong from the ship. Over the years, Mr. Snow managed to keep the saga of the *Portland* in the public's mind. He wrote and theorized about it in numerous volumes about storms of New England. Mr. Snow did 35 years of research and interviewed many of the principals who figured prominently in the story. Some of Mr. Snow's friends also did many hours of research for him on the *Portland*, both in Massachusetts and in Maine.

In October 1998, my wife and I spent a week at the beautiful Samoset Resort in Rockland, Maine. I spent 60 percent of this time doing research for this book. In looking for the slipway that the steamer had been launched on in 1889, I was told by locals that the slipway was long gone. The yard had closed in 1906 and a steel yard, The Texas ship-building Company, had been built in its place. I was told that docks had been built across the location of the slipway. In working with Mr. Snow, he taught me never to give up in looking for something and to look at a site in two different ways, once at high tide, and once at low tide. At high tide, it looked just as described -- old foundations, meadows, and trees. Four days later, on a return visit at low tide, on the mud and sand bottom of the Kennebec River, I found the slipway that was not supposed to exist.

In the Edward Rowe Snow archives at Boston University, there is a photo from the Smithsonian Institution that shows the New England Shipbuilding Company in Bath in its entirety. The photo was taken from across the river in nearby Woolwich. At the southern end of the yard, there is a large white steamboat on the ways shortly before launching. In the back slightly to the right, there is a house on Front Street with three white chimneys with black tops. That house is still there, and was used to pinpoint the exact position in the yard where the *Portland* slipway was located. In comparing this with close-up photos of the *Portland* stern at dockside, counting the portside cabin windows from the paddle box back, matched the number of the windows on the *Portland* in the Smithsonian photo.

The only other steamboat being worked on at the time was the *Kennebec*, which had windows paired, not singularly as in the *Portland*. The date of the photo is actually 1889, not 1890 as in the photo caption, as the *Portland* was launched on Oct. 14, 1889 and immediately towed downriver to the Bath Iron Works to have its boilers installed.

Going to the exact location at low tide, I found two 20-inch diameter logs sticking out of the riverbank at a downward angle, about 15 feet apart, with one of them notched out on top with multiple saw cuts to attach the next section to it. I then photographed it from four different angles while shuffling through the mud of the Kennebec River in my snow boots. I then took the film to a local film lab and had it processed in one hour. Later, I showed them to the people that told me it was long gone. The look on their faces were priceless – so much for impossible!

Since early colonial days, the beaches of outer Cape Cod have been the final resting place for more than 2,000 ships. During any northeast storm, as the outer beach is downwind, the ships that do not have enough power to beat into the seas are tossed onto the beach or outer reefs. This usually kills a large percentage of the crew, who drown in the raging surf. One of the most famous of these was the pirate galley *Whydah,* captained by Black Sam Bellamy. Aboard her was treasure from more than 50 ships and it has lain in the sands for 300 years. After many years of diligent research, treasure-hunter Barry Clifford of Provincetown located the wreckage of the pirate ship in less than 20 feet of sand off Wellfleet. Since he had enlisted the help of famed treasure hunter Mel Fisher of Florida, who found the location of the 1600s Spanish treasure fleet (most notably, the Atocha), the Massachusetts State Board of Underwater Archeology said publicly that Clifford had "salted" the site with some of the gold coins from the *Atocha*.

If he had, the coins would have been later proven to be of Spanish origin from the 1600s when in fact, the *Whydah* sank in 1717, only a year after it had been built in England. In October of 1985, during one of the last dives of the season, Clifford's boat, the *Vast Explorer,* had just blown a hole 20 feet deep and eight feet wide in the sandy bottom, and sticking out of the side of the hole was the definite lower section of large bell. Divers quickly enlarged the hole and the bell was hoisted out of the water and swung aboard with a large winch.

Most of the marine growth was carefully cleaned off, but lettering on the top of the bell was covered by a thick concretion that would not yield. The bell was taken ashore to the salvors headquarters in Orleans and put into a large tank of fresh water. The next day the concretions at the top of the bell were very carefully chipped away by the lab conservators to reveal the words "The *Whydah Gally* 1716." So much for salting the wreck site! The Board of Underwater Archeology became quite embarrassed over the incident, as it had said in print in the local papers that the site was unequivocally not the *Whydah*.

Since then, more than 400,000 artifacts have been found preserved and catalogued, while none of it has been sold to cover the costs. Today, one can see displayed at the *Whydah* Museum in Provincetown the bell and many of the salvaged artifacts. Over the years, as soon as a shipwreck hit the beach and any survivors were rescued, locals would descend on the wreck and start picking it apart. After the *Whydah* was wrecked in 1717 with very few survivors, the Commonwealth of Massachusetts sent Captain Cyprian Southack down to the outer Cape within days to seize the ship's treasure for the Commonwealth. By the time he got there, the locals had picked the beach clean.

Everything of value had already been hidden away in homes, cellars, and barns. When Captain Southack raised a ruckus about this to the local authorities, some beat-up pieces of the wreckage were surrendered by the locals. It cost more to journey to the Cape than any value recovered.

Later, during the early 1800s, local thieves known as Mooncussers would set up a fake light on the beach to look like a lighthouse to cause ships to wreck. These dastardly men had no intention of rescuing anyone from the ship and would just as soon kill them if they made it ashore alive. In the business of stealing cargo for later resale, they did not want any witnesses. This went on for years until the start of the U.S. Lifesaving Service nightly beach patrols, where some of the Mooncussers were almost caught red-handed.

During a storm, a ship did not even have to hit the beach or an offshore sandbar. Sections of many previous wrecks were sticking up off the bottom, and a deeply laden ship could also hit one of them and rip its bottom open and sink off the beach. Hopefully, help would come from ashore before crewmembers froze to death in the rigging. This happened to thousands of unfortunate sailors over the centuries. Over the centuries, over three-thousand ships have been wrecked on Cape Cod alone.

In looking for wrecks on the outer Cape, treasure hunters must tow a magnetometer behind a boat and cover a large area. Over a period of time, several "hits" are made, some being much stronger than others. In evaluating the "hits", it is very difficult to tell if the hits are from a steel (recent) wreck, or from cast-iron cannon located deep under the sandy bottom. Side-scan sonar also towed behind a boat is later brought over the positions of the hits and the composition of what is below the bottom can be studied from a computerized photo. The sands off the Cape shifted so much that just after a storm is the best time to look. Gold and silver coins from long ago washed up on the beach at these times.

The treasures and artifacts are now housed at the National Geographic Society Museum in Washington, D.C. as well as the Pirate Museum on Macmillan's Wharf in Provincetown and in the lab in Orleans where many more artifacts are undergoing restoration.

After December 6, 1898, no more bodies of the *Portland* victims had ever come ashore. The last victim had been picked up heavily sanded in near Chatham. The rest would have remained buried in the sand or were carried along the bottom out to sea, south of Monomoy Island and lost forever. The sea does not give up its secrets easily.

CHAPTER 12

The Storm Still Rages

OF ALL THE HARBORS ON the Massachusetts coast, Scituate Harbor on the South Shore bore witness to one of the most heartbreaking scenes.

On their way to Boston from Norfolk, Virginia, under tow by the towboat *Mars*, were the barges *Daniel Tenney* and *Delaware*. The three vessels were slammed by the storm while off Minot's Light, and the *Mars* was forced to cut both the *Tenney* and the *Delaware* loose to anchor and fend for itself. The *Delaware* sank off Collamore Ledge near the Cohasset-Scituate border, killing the captain and her five-man crew. Parts of the wreck of the *Daniel Tenney* were found on the beach by the surfmen of the Fourth Cliff Lifesaving Station. None of the crew's bodies ever were found. At that time a sand and stone beach existed across the gap between the Third and Fourth Cliffs in Scituate. The surfmen normally crossed this stretch of barrier beach on the northern part of their nightly patrol. Over the years, shipbuilding had declined on both the North and South rivers between Scituate and Marshfield because the southern and very narrow entrance of the rivers filled in with silt in the days before dredging.

The entrance became so shallow there wouldn't be sufficient depth at high tide to get out of the river in the event bigger ships were built there. By 1898, most shipbuilding on the rivers was non-existent. During the storm, the huge seas began narrowing the width of the barrier beach. At the same time the huge tidal surge had built up in the marshes behind the beach. This combination caused the volume of water behind the barrier beach to take the path of least resistance back to the sea and completely blew out the entire beach. Where the surfmen from the Fourth Cliff Lifesaving Station had walked this strand of sand and shale, only a few hours prior to this disaster, a huge gap now greeted them that was half a mile wide and more than 20 feet deep. This immediately cut off the lower section of Scituate at the Fourth Cliff area known as Humarock from the rest of the town and did almost irreparable damage to the Fourth Cliff Lifesaving Station. The shipbuilding problem had been solved by nature. What man couldn't do, nature did instantly.

Residing at Scituate Third Cliff was Captain Frederick Stanley, who after years on the sea aboard merchant ships, joined the Lifesaving Service in 1880 shortly after his retirement so that he could settle down and raise a family. As Captain Stanley's work as a surf man was so exceptional, he was later promoted to the head lifesaver position at the Fourth Cliff Lifesaving Station.

One of the priorities of the surfmen during severe storms was to warn the temporary inhabitants of nearby gunning shacks set up in the extensive marshes between Scituate and Marshfield to the hazards of being trapped in the stormy tidal surges. In some cases, the surge could rise 10 feet above the mean high-tide level, completely engulfing anything in its path.

Surfman Richard Wherity usually carried out this vital task of checking the marshes behind the station while most of the other surf men were out on beach patrol. In the excellent book "Warnings Ignored," writers David Ball and Fred Freitas tell the story of the two Henderson brothers from Norwell and hunting friends Albert Tilden, George Webster, and George Ford. Out in the storm were some other buddies staying at their hunting shanty on the South River, the Clapp brothers, Richard, William and Everett.

They were out hunting for birds during the post-Thanksgiving holiday weekend. Surf man Wherity took one of the station's small boats and rowed out into the marsh to warn the both groups of hunters. On arrival at the

Henderson shanty, his warnings were brushed aside by the Henderson brothers, Tilden, and George Ford. Webster gathered his gear and headed back with Wherity in the boat.

They next stopped at the Clapp family shanty and Wherity's offer also was refused by the entire Clapp party. Both Wherity and Webster then rowed back to the Fourth Cliff Life Saving Station and relative safety. With the sudden intensity of the storm, the Henderson party decided to immediately abandon their shanty behind Fourth Cliff in the South River and make for shore. Upon going outside, they were horrified to find their gundalow missing. A gundalow is a shallow drafted type of cargo barge, once common in the Gulf of Maine's rivers and estuaries.

At approximately the same time, the Clapp party decided to abandon its camp in the South River near the entrance to the Snake River. Their small boat with three adults in it was being tossed all about and taking in water from the raging river. As their boat was about to go under, the large gundalow, which had broken away from the Hendersons' shanty, floated by. They immediately rowed over to the gundalow and jumped aboard with all of them making it. In the frenzy to grab the wayward gundalow, the rowboat was lost, as well as its oars. Since there were no oars in the Henderson's' gundalow, they were carried helplessly southward toward Marshfield, where the barge came to rest in the marshes.

The Clapp brothers headed for the nearby Marshfield Hills railroad terminal to get warm before starting the long trek home via Norwell because of washed out bridges and roads. All of the Henderson party that had decided to take a chance by ignoring Surf man Wherity's warnings, perished in the tragedy, their bodies being found ashore over the next five days.

In North Scituate Sand Hills section, many houses were blown off their foundations by the huge seas and landed upside down either across the street or in the salt pond on the marshes inshore. Wreckage from houses, ships and barges that had sunk off the coast was all mingled together, along with tons of coal which had been on the wrecked barges. The residents who later gathered up this wreckage would have coal and firewood for heat for months.

Two vessels that had anchored in the harbor had dragged ashore. The saddest loss of life occurred on the beach at Sand Hills. According to Edward Rowe Snow's 1943 volume, "Storms and Shipwrecks of New England," "The first of the many wrecks caused by the Portland Gale was that of the pilot boat *Columbia*. The *Columbia's* last known act was to discharge her last available pilot, Captain William Abbott, aboard the steamer *Ohio*, heading into Boston.

The storm grew in intensity as the storm wore on, and it is believed that the *Columbia* was in the vicinity of the *Boston Lightship*. No one escaped the wreck, so conjecture alone must dictate the story of the loss of the *Columbia*." The Pilot Boat Number 2 had been built in Boston only three years before and was a fully seaworthy vessel and fully manned. Prior to her loss in Scituate, she had been out at sea putting harbor pilots aboard at least four vessels. This had to be done when the incoming ships are far off the harbor, so that the pilot boat had to be out at sea at least 15 miles as the *Boston Lightship* lies 13 miles to sea out from Boston.

"The following morning, Surf man Richard Tobin of the North Scituate Lifesaving Station had the south patrol along Sand Hills from 8:00 a.m. to noon. His report says, 'I went down to the beach to the key post about three miles from the station. The seas were coming over with such force that I was washed into the pond back of the ridge. It was blowing so hard that I was obliged to kneel down to get my breath,'" Snow reported. At 9:30, Surf man Tobin was on the veranda at the very cottage into which the *Columbia* crashed later in the day. When he returned to the station it was three in the afternoon.

"The storm was now so intense that no patrol was permitted to leave; in fact, not until midnight, did the next patrol go out. Surf man John Curran, set out on the south patrol this time. At 1:45 in the morning, he sighted a schooner right in the line of his patrol, lying on the beach where she had crashed into a cottage.

"The wreckage from the shattered building (owned by Otis Barker) was lying in heaped confusion on the decks of the careened vessel. Since Surf man Tobin had stood on the veranda of that very cottage at 9:30 Sunday morning, the *Columbia* had washed up on the beach between then and 1:45 a.m. on Monday. No one witnessed the disaster and no one ever knew what took place. By 3:20 that morning Curran had returned to the station and notified the keeper, who visited the scene with three surfmen. On the way, they found the body of the *Columbia*'s first boat-keeper. The other four men were later found dead in the vicinity. The wreck of the *Columbia* was complete. The starboard side of the pilot boat had split near the garboard (first plank out from the keel), the planking was torn where she had hung and ground on the rocks, and the sternpost was broken to pieces. The foremast was gone, two anchor chains hung from the hawse pipes, and both of the anchors were missing.

"Probably the five men on the pilot boat were dead before the *Columbia* struck the beach and the fact that both anchors were gone showed that the vessel had unsuccessfully tried to weather the storm. This opinion was shared by a majority of seafaring men who visited the scene."

Upon seeing what had happened to his cottage, owner Otis Barker made the wrecked ship into a museum and a tea house. He made money from this for 30 years, until the house and ship were threatened with imminent collapse. In the early 1930s the Town of Scituate ordered both the dwelling and the wreck to be burned to rid the beach of the hazard. A fireplace mantle from the *Columbia* was inscribed with the words "Glad to see you. You'll find us rough, but you'll find us ready." It now graces the fireplace in the Maritime and Mossing Museum in Scituate.

How much of the inscription was from the *Columbia*'s crew and how much of it was from Otis Barker remains a question for the ages. In Mr. Snow's words, "Thus ended the career of the Boston pilot boat *Columbia*, the schooner on which five brave men tried unsuccessfully to ride out the Portland Gale."

After the new opening to the North River changed things forever, Fourth Cliff in Scituate has to be reached by going through neighboring Marshfield, and up to an area of Scituate known as Humarock. This opening remains one of the most treacherous river mouths on the on the south Coast of Massachusetts. The tugboat *Mars* later made Boston Harbor, showing much damage from the storm. On Tuesday, Nov. 29, 1898, the schooner *Hiram Lowell* sailed into Gloucester Harbor with the survivors of the British schooner *Narcissus* aboard.

The chief officer of the Narcissus, Captain McIntosh, tells the tale: "The schooner left Boston Wednesday evening with a cargo of 1,120 barrels of flour, and passengers bound for Liverpool, Shelburne, and Lunenburg, Nova Scotia. She stopped the next day at Gloucester, and took on more passengers. She cleared from this port Friday afternoon at 5:00 o'clock Saturday evening when about 70 miles west of Seal Island, she encountered a tremendous gale. Sail was quickly shortened and until Sunday morning she ran under a three reefed foresail. At that hour the little sail was torn to ribbons, and the vessel was then at the mercy of the waves, which made a spontaneous break over her. At about 11:00 o'clock, the "jumbo" and jib were carried away, together with much of the headgear.

The vessel was found to be leaking badly, and all hands were called to the pumps. So great was the leak that the water was found to be gaining against the combined efforts of crew and passengers. Then too, one of the pumps broke down and refused to do its duty. Those aboard the ship abandoned hope but some worked on. Others prayed while the waves broke over them. All through the night the water continued, to gain, until she was thoroughly waterlogged. When morning dawned, there was no sign of rescue or help, still, signals of distress were hoisted and the men worked on at the pumps.

At 3:30 p.m. Monday afternoon, when 40 miles east by south of Highland Light in Truro, a vessel was sighted which proved to be the *Hiram Lowell*. The latter sighted the *Narcissis* and sailed up to her and stood by.

About the same time, the ocean liner *Philadelphia* hove into sight, and seeing the signals of distress, ran down to the *Narcissis* and offered to take her in tow, but the offer had to be refused for the fear that the strain would tear her apart. The schooner was then abandoned and the survivors taken aboard the *Hiram Lowell*, which thanked the *Philadelphia* and immediately headed to its next stop in Gloucester."

As seen in the story of the rescue of the crew of the *Narcissis* without any loss of life, luck and heroics played a large part in the crew's survival. Had the crewmembers not kept to the pumps, long after the passengers had given up, the story would have ended quite differently, with everyone lost. The captain of the *Hiram Lowell* and his crew should have been awarded medals for their actions. After its shameful act in cutting both *Barges numbers 1 and 4* off Point Allerton, the tug Underwriter headed out into Massachusetts Bay to get some offshore maneuvering room. Later, arriving in Boston, the *Underwriter* crew-members told their story to a Boston Herald reporter.

"After cutting loose the two barges, we headed out to sea and have been cruising around the Bay ever since. Sometime Saturday night or early Sunday we lost our second and last anchor and 150 fathoms of cable. One of our boats was smashed and our main rigging parted. Four hatches were blown off and everything below was awash. About 10 a.m. Sunday the forward hatch was torn off, and we were obliged to put a drag on, for about two hours, to keep the boats head up (into the wind) while we fixed it. The windows of the pilot house were broken, and the captain's room was badly damaged.

Sunday afternoon, the main pump gave out and we had to bail out the fire room every other hour. Not a man on board had closed his eyes since Saturday noon, and not one has had a square meal. It was impossible to cook so we lived on bread, hardtack and canned goods until today. The tug will need repairs, as she is in bad shape below. Everything has been covered with salt water for 48 hours. The engine room especially shows signs of the storm, as all the metal work is rusted by the water which came through the broken hatch. New anchors and cable are needed to replace those lost, and new rigging for her mainmast. Days later when some of the rescued crewmen from *Barges 1 and 4* came aboard the *Underwriter,* they were just about ready to kill the captain, they were so mad."

Off the coast, nothing had been heard from the four-masted schooner *King Philip,* which had last been seen off Portland. According to the Boston Herald of November 30, 1898, "The Schooner King Philip of Fall River has been totally wrecked on Cape Cod near Highland Light.

"As wreckage from her has come ashore, the ship had been built in 1888 in Camden Maine, and hailed from Fall River. She was well known in the coastwise trade, and was one of the best of the older vessels in her class. She had weathered many a hard gale, but always managed to get out with comparable little damage. She was a splendidly built craft and was considered staunch and speedy. She was 211 feet long, 42-foot beam with a draft of 20 feet. The crews of approximately 10 were all considered lost. Three days after the storm ended a large pump still fastened to a section of her deck was found inside Cape Cod Bay with the name of the ship on it."

Several fishing schooners had left Gloucester days before the storm to fish on Georges Bank. One, the *Gloriana,* had been fishing on the Grand Banks as well as George's Bank, was reported to have been lost off Race Point Provincetown.

The Boston Herald noted that "The *M.H. Walker* also reported wrecked off Race Point, hailed from Gloucester. She was built at Essex in 1889, and owned by Edwin McIntyre. The *Walker* was engaged in the winter haddock fishery and was commanded by Captain Frank Miller, formerly of the schooner Norman Fisher. She had carried a crew of 14 men. Captain John T. Denon of the sloop *Venus* and his crew, reported missing at Plymouth, are all safe and well. The Venus went ashore on Clark's Island, off Plymouth during the Gale, and is probably not badly damaged. Judging from the report of Captain George Lewis of the schooner *New England,* who arrived today, however, the Gloriana could not have been in the vicinity where it was stated she was wrecked. Captain Lewis says that the *Gloriana* left here last Wednesday in company with him for George's Bank.

They arrived Thursday night. Friday was unsuitable for fishing and Saturday the first set was made. Captain Lewis could not understand why the *Gloriana* could be in the locality it was reported she was lost, as he supposed at the time she was on George's Bank. Captain Lewis reports that the schooners *Edith Wilson, Whal*en and *Mizpah* were also fishing on George's Saturday.

New England had the gale heavy and drifted from the Northern part of George's to the south channel, where she sighted a coaster on fire yesterday morning. The only damage which the *New England* sustained was a broken main boom, the work of the sea.

The schooner *Shenandoah* arrived in Gloucester from the Grand Banks this morning. She came out of Pubnico, Nova Scotia, Saturday and encountered toe full fury of the gale, having dories smashed and all boats washed from the decks. Sometimes the only remains of a lost ship are a message written on paper and cast adrift in a sealed bottle.

Such is the story of the rescue of the crew of the schooner *Emma Dyer,* of Gloucester by the steamer *Herman Winte*r of the Metropolitan Line, running from New York to Boston. The schooner had been terribly knocked about by the elements, and that her crew felt there was no hope of life was shown by the following letter picked up in a bottle near Nauset Beach in Orleans, Cape Cod:

Our spars and cables are all gone. We are drifting ashore in a living gale of wind. All hands expect to go; God save us all. Have mercy on our poor souls. My name is Neil McNeil. I leave a wife and child, 93 Maplewood Avenue Gloucester.

Anyone who picks up this bottle, report by wire. We are 16 in crew; my dory mate is Peter Hovel. God have mercy on us all, Schooner *Emma M. Dyar*."

When McNeil wrote this, he doubtlessly felt, as did his crewmates, that they were face to face with a supreme crisis of life. For 36 hours they had been buffeted by wind and storm, and their little vessel so badly strained that she seemed destined, beyond all doubt, to go the bottom. However, something marvelous happened.

Just as the captain and the crew were just ready to give up all hope, the *Herman Winter* providentially hove into view, took them all aboard by throwing ropes to them, and then taking their schooner in tow, carried them all to New York, so that so that if anybody obeys a Mr. McNeil's injunction he ought to do so with the understanding that Mr. McNeil is safe and sound instead of being at the bottom of the sea as he expected to be when he wrote this letter.

The three-masted schooner, *Edgar Hanson*, coal-laden, bound for Portsmouth, New Hampshire, went ashore on Dread Ledge off Nahant during the storm. The vessel was discovered blowing horns for help in Nahant Bay. She lies in a dangerous position and if the wind changes, she is liable to be battered to pieces on the rocks. The Swampscott Lifesaving crew under

Captain Horton has gone to her assistance. She carries a crew of six men." The rescue attempt was successful, but the ship was a total loss.

Reports from Mattapoisett, today, stated that only one man of the schooner *Hattie A. Butler* which went ashore on Angelica, Buzzards Bay, and early Sunday morning was drowned. Captain Peter Mullen and a foremast hand succeeded in reaching the shore in the raging surf, but Thomas Atkins of Brooklyn New York was drowned in the attempt.

All up and down the coast, eerie harbor scenes were the same; wreckage, chaos and death. There were a few heroes mixed in, too, that kept the death toll much lower than it normally would have been. Everywhere one traveled, onshore or tried to, they were met by one obstacle after another.

Railway tracks were twisted into grotesque shapes, railway beds totally washed out and trees and telephone poles blown across the tracks. Any kind of transportation was slowed considerably.

In Fall River, the tug *Concord*, which sailed Saturday night for Philadelphia with light barges *Satin Ella, Taunton, Pioneer* and *Star of the East,* broke adrift near *Hog Island Lightship* and the *Pioneer* and *Star of the East* went hard ashore on Prudence Island. The schooner *Nellie W. Craig*, which arrived here from Newport News, November 28th, was blown ashore during the gale.

Off New Haven, Connecticut, on Nov. 29, the tugboat *Carbonero*, of the Boston Towboat Company, which reported its three barges lost off Oilfield Point, Long Island put into port here tonight and reported that she found her three barges ashore with the crews all safe. On the same day in New Haven Connecticut, the Tugboat *Herald*, which had cut loose three barges, have not yet been heard from? They are the *Escort* and the *McCauley*, both of Boston, which with the Naversink composed the tow of the tug *Herald*, under the command of Captain Hersey of Boston."

CHAPTER 13

Filming the Sunken Portland

—

Dr. Craig MacDonald, formerly of the University of Hawaii, moved to Massachusetts to become the new director of the Gerry Studds Stellwagen Bank National Marine Sanctuary. He had not heard of the story of the sunken Steamer *Portland*, so I helped bring him up to date on the wreck that lies within the boundaries of the sanctuary. The wreck had been discovered in 1989 by four of the best underwater wreck locators in the world, Arnold Carr, Peter Sachs. Herbert McElroy and John Fish of American Underwater Search and Survey. Their company had been hired by numerous federal agencies to find the remains of airliners that had crashed at sea.

On their own, over a 20-year period, they had sought the wreckage of the *Portland* for a book they were writing about the ship. After scouting many wrecks around Cape Cod, where the *Portland* was supposed to have gone down, using the latest technology, no wreck that they had looked at even had the characteristics of the ship. For one would have to show the two most identifying traits of the ship, the paddlewheel boxes and the two twin side-by-side smokestacks.

When a wreck was located about 20 miles north of Cape Cod, researchers took underwater photos and announced in 1989 that they had found the *Portland*. According to the Stellwagen Bank National Marine Sanctuary officials, "In late July and early August, a Joint Research Mission involving both NOAA's Stellwagen Bank National Marine Sanctuary and the National

Undersea Research Center at the University of Connecticut, mapped and shot video of the wreck lost in the Portland Gale of 1898. The video and side scan images from the mission provide visual documentation to earlier work done by American Underwater Search and Survey."

"We are excited to be able to bring some closure to one of New England's most mysterious shipwrecks" MacDonald said. "The story of the steamship *Portland* and its fatal last run from Boston to Portland, Maine has intrigued maritime historians for years due to the wide-ranging reported sightings of the ship during the storm. This mission allows us to start putting some answers to the questions about its loss. Side-scan images from the research vessel RV *Connecticut* and the NOAA ship *Ferrel*, showed that the wreck sits upright on the sea floor, with its hull largely intact but much of its superstructure gone," said primary investigator Ben Cowie Haskell of the sanctuary. Wreckage from the vessel found along Cape Cod beaches in the days after its loss, included primarily pieces from its upper decks. "All passengers and crew were lost, but the exact number has never been determined due to the lack of a passenger list on shore," said Haskell. "The latest estimate is 192 individuals lost, with only 38 bodies recovered s they washed up on the Massachusetts beaches between Truro and Monomoy.

"Remotely operated vehicle operations from the RV *Connecticut* in July produced high quality video footage of the wreck that showed the steam release vent, rudder assembly, paddle guard, paddle wheel hub and overall length. The observation of these features positively identified this wreck as the *Portland* as there are no other coastal steamers of this type reported to have been lost in Massachusetts Bay. Although artifacts displaying the ship's name could not be found, a team of independent researchers confirmed the identification based on the evidence provided by the side-scan and video images.

"Aside from the overall size and shape of the explored wreck, one of the most compelling pieces of evidence to support the conclusion that it was the *Portland* was the spittoon-shaped end of the steam escape pipe found twisted and lying on the seafloor at the site. It matches the escape pipe in the Antonio Jacobsen painting of the steamer."

Along on the survey were Dr. Ivar Babb from NURC at the University of Connecticut and the young students were from the University's Aquanaut programs who are studying to be oceanographers. One of the teachers of the program, Kathryn Zubrowski, found the "match" in matching the spittoon-shaped end of the steam escape pipe found on the ocean floor near the wreck to its position just aft of the twin smokestacks in the old Antonio Jacobsen painting provided for research purposes by the Maine Historical Society. The *Connecticut* had aboard on the expedition trip out of Gloucester both the side-scan sonar and ROV vehicles, and was one of the smallest ships in the world to have the Dynamic Positioning System aboard. This computer-controlled system automatically keeps the ship in position over a wreck without having to use anchors. In deep water, it is a godsend.

A research fellow from the University of North Carolina, Matthew Lawrence, also was there to plot the lines of survey similar to cutting grass in the ship's attempt to "mow the lawn," dragging the side-scan sonar on multiple passes behind the ship.

The Connecticut was under the able command of Captain Turner Cabiness. "As the ship passed over the location we carefully watched the ship's Fathometer as it dramatically changed depth by over 40 feet. There was obviously something big down there," said Turner, as his assistant student, Amberlie Silva, recorded the exact latitude and longitude. Following this initial confirmation, the ship headed for the first of four side-scan sonar lanes set up by Lawrence. The monitor instantly became the center of attention as we began to run along the survey tracks. The first pass produced nothing out of the ordinary, other than evidence of nearby fishing activity on the seafloor. On the second lane, however, a distinct shadowed object began to roll down from the top of the computer screen as the vessel ran its course. "There she is!" exclaimed Bruce Terrell, one of the oceanographers aboard, and we all watched as the entire screen filled with a clearly man-made object. The next pass produced another striking image of what clearly was a wreck, but the burning question remained unanswered: "Was it the *Portland*?"

"I suggested that since the ROV was only on the ship for these two days, and it was the tool to unequivocally identify the wreck, that we forego additional sonar surveys for the day and begin diving operations. The ROV crew quickly swung into action as Turner initialized the dynamic positioning system over the wreck. The first dive was the first of a three-act drama of discovery. The ROV descended to the seafloor using its sector scanning sonar to be able to pinpoint the range and bearing to the large wreck. Archeologists and Sanctuary SRC Personnel crowded into the NURC-Nagl's ROV control van that is setup with computer screens for the integrated navigational system, the sonar, plus video screens for the two video cameras mounted on the vehicle.

The van also has an additional TV monitor and computer video capture workstation for scientific use. The dry lab on the ship was dedicated to the sonar computer and an additional TV monitor connected to the ROV that served as the biological observation station. Finally, Sanctuary educational personnel were carefully watching an additional TV monitor in the vessel's galley wired to the ROV output to show the teachers and the ship's crew the happenings down below. Approaching the wreck, we encountered a few pieces of debris that provided clues that we were getting close. Then suddenly, the video screens filled with a solid vertical wall, the side of the wreck. As NURC'S ROV pilot, Craig Bussell, began to slowly maneuver the vehicle up the outward curving hull, the screen was filled with the bright oranges, pinks, and whites of sea anemones attached to the wreck.

Then all the screens went black. The ROV had lost power and had to be recovered by hand as everyone's mind and mouths raced with questions, guesses, supposition and hypotheses as to what was wrong with the vehicle.

Was this a fatal breakdown and what had we just seen? NURC's team quickly began to troubleshoot the problem, while the side-scan sonar crew spun up to conduct a survey of another significant wreck site that lay only kilometers from the *Portland* site, but that was another story."

"Act two of the drama ensued in the afternoon, as the ROV was quickly repaired. Again, the ROV descended from the dynamically positioned ship; this time, from a slightly different location over the wreck. As the ROV ran along its length, everyone wondered what this pipe as that lay solitary on the seafloor with its base bent over at a 90- degree angle. The vehicle encountered the large hull of the wreck and began a detailed survey that produced numerous clues that this was indeed the *Portland*. As the ship headed to port at the end of the day, the tapes were being reviewed. It was this time when the spittoon-shaped end of the steam escape was found matching the old painting of the ship and its uniqueness matched it to the long-lost *Portland*.

The final act of the drama played out the following afternoon, following additional side-scan sonar surveys of the *Portland* wreck site. The ROV descended on its third dive as the vessel was positioned over the stern of the *Portland*. The dive began without incident, but blood pressures began to ascend in direct relation to the number of lost gillnets the ROV encountered as she worked along the stern of the vessel. The rudder was clearly documented, and then the vehicle tried to proceed forward only to come into contact with the lost nets, a feat that experienced ROV pilots know well, rearing up like a lassoed horse or a dog on a leash, indicating that the tether was caught. Numerous attempts to pilot out of the snag proved fruitless.

Meanwhile, the *Connecticut* remained rock solidly positioned over the wreck. Had it not been for this capability in the face of a fresh breeze, the ROV probably would have been lost. In the end, finesse gave way to force and the ROV was retrieved by the ship's winch, along with one hundred pounds of lost net and settled fauna. All breathed sighs of relief. The epilogue of the drama was fittingly a few moments of silence and words of remembrance for those lost on that fateful day of Nov. 27, 1898: "May this be forever a sanctuary for their souls," MacDonald said solemnly. Three bunches of brightly colored flowers were cast upon the water, the aquanaut students and slowly drifted away from the ship during a moment of silence. We then returned home to reflect and to plan for the next year".

CHAPTER 14

New Developments

———

SINCE 1989, WHEN JOHN FISH and Arnold Carr would venture out to the wreck of the *Portland* to investigate, they would hire a 60-foot dragger out of Scituate to have a steady platform to lower underwater cameras to image the wreck. Three days of calm seas would be needed to position the trawler precisely over the site in more than 400 feet of water. By dropping three 6,000-pound anchors over a period of days, the boat could be shifted in position over the wreck to get the best photos. An underwater camera suspended from a long line in the currents could not be precisely controlled and positioned like Remotely Operated Vehicles, which can be had for only $250,000. In addition, the wreck is partly covered with old fishing nets and the cables that go with them. In the 100 years since the ship sank, very many fishing nets have snagged on the wreck and only add to the danger of losing an ROV. If a manipulator arm is attached to the submersible, it can cut its way out of the nets. However, sometimes, the arm is not used. Instead more lights are put on the ROV to obtain better photo images.

In July 2002, the federal government commissioned a survey of the underwater resources inside the new Stellwagen Bank Marine Sanctuary. This included the shipwrecks on the bottom. Prior to this search, officials from the Marine Sanctuary contacted the HMG Group at American Underwater Search and Survey in Bourne and asked them about the location of the Steamer *Portland* on the bottom of the northeast portion of the bank. As they no longer had the time to image the wreck and could not afford their own ROV, HMG graciously gave the federal officials their coordinates for the location of the *Portland*, and agreed to share information about the wreck. It was then found easily with side-scan sonar and an ROV was lowered more than 400 feet to image the ship.

For the first time in history, the *Portland* was seen in awesome detail, as underwater imaging had grown better in leaps and bounds. Only the lower part of the hull is still there, as all the upper works (superstructure) was ripped off during the sinking. MacDonald said the team that videoed the wreck was able to conclusively prove the wreck as the lost steamer *Portland*. To prove this, they showed that the twin side-by-side smokestacks were still on the ship. This greatly surprised me, as I had thought that the smokestacks would be the first things to go during the sinking. Both boilers appear to be intact, as is the 50-foot-high A-frame that encompasses the massive 2,000 HP engine made by the *Portland* Engine Company. On the top of the A-frame still attached is the 22-ton walking beam. The ship's rudder, which is still on the wreck, was videoed without a propeller. The combination of the two side-by-side smokestacks, with the walking beam engine and the now missing steel paddle boxes, is irrefutable proof that this wreck could only be the *Portland*.

On Aug. 28, 2002 in Braintree, I lectured on the loss of the *Portland* for the Massachusetts Bay Fishermen's Association. The very next day on the noontime news on WCVB Channel 5, I saw the announcement that the wreck of the *Portland* had indeed been found. The underwater video was shown for the first time, and it was brilliantly clear. The images showed the paddle-wheel guards as well as the ship's old-fashioned folding stock anchor still attached to the wreck, as if it had sunk just last week.

More research and video will be conducted on the wreck in the future, and the exact depth and coordinates of the wreck have to be kept secret to keep scavengers from stripping artifacts and archeological information from the ship. With NOAA's direction and participation, artifacts may eventually be brought up from the wreck to be conserved and put on display in a museum for the entire world to see.

Conserving artifacts raised from the bottom of the ocean is a very expensive business and has to be done just right. The artifacts are immersed in a large tank of freshwater and electrical wires are attached. A small electrical current passes through the wires, and the slow electric action removes the salts from the metal stabilizing it. Over a period of time, from months to years, depending on the size of the artifact, the salts in the metal are removed and the object is removed from the tank and placed in a glycol solution to finally preserve it. After removal from this step, the glycol is allowed to dry in controlled conditions, after which the artifact can be put on display for the public to see.

CHAPTER 15

Myths and Realities

—

Myth: The Boston agent, Charles Williams ran down to the ship to stop Captain Blanchard from sailing.
Reality: He never ran after the captain. The Boston agent had the authority to only relay commands and suggestions from management in Portland. The captain had his standing orders to sail at 7 p.m. The final decision to sail was the captain's alone, unless he was ordered not to sail by upper management.

Myth: It was snowing when the ship left India Wharf.
Reality: The snow did not start until 7:30 p.m., at which time the *Portland* was steaming through President Roads in the outer harbor.

Myth: Just before the sailing time, a black mother cat was observed removing its litter from the ship and depositing them on the dock.
Reality: This may have been so in a small way, but it has been retold and embellished so many times over the years it is impossible to define where the exact truth lies.

Myth: The *Portland* was never seen again after she left the harbor.
Reality: She was seen from many other ships and from shore after she cleared Boston Harbor, and from Thatcher's Island Light as she passed there after having completed the first third of the voyage. More ships out in the storm also saw her after she passed Thatcher's Island before she foundered.

Myth: Captain Michael Francis Hogan, skipper of the schooner Edgar Randall, saw the *Portland* off of Provincetown on Sunday morning.
Reality: It would have been impossible for the *Portland* to be seen off the tip of Provincetown at approximately 7 a.m. on Sunday morning and then to have steamed into the 30 to 40 foot seas more than 24 miles in two-and-a-half hours to sink off of Gloucester at 9:30 a.m. No ship at the time had the visibility or horsepower to do that. What Captain Hogan actually saw was the salt-and-ice covered steamer *Horatio Hall*, hove-to off of Provincetown. Her black steel hull would appear white, the same color as the *Portland*, and three quarters down the side of the ship would be hump-shaped wave washout that appears black. This, from a distance, would look like a paddle box, thus explaining the discrepancy.

Myth: The captain disobeyed manager Liscomb by going to sea.
Reality: Liscomb made the suggestion through agent Williams that Captain Blanchard should hold the sailing until 9 p.m. to gauge the weather, and if it looked threatening, to cancel the voyage. Liscomb testified during the later inquiry that he did not make it an order. The Captain acted on the latest weather information available and chose to sail at the regular hour.

Myth: The four short blasts heard by Captain Fisher of the Race Point Lifesaving Station came from the *Portland*.
Reality: At the later inquiry, Captain Albert Bragg of the *Horatio Hall* testified that his vessel was hove-to off Provincetown, but was never in an emergency and did not sound four short blasts, a recognized emergency signal. That signal must have most likely come from the 128-foot screw-steamer *Pentagoet*, also lost with all hands off the Cape.

Myth: Against all laws of the sea, Captain Hollis Blanchard should not have sailed and exposed his vessel that night.
Reality: In later interviews in his research on the loss of the *Portland*, Edward Rowe Snow discussed this with 13 retired sea captains at the Boston Marine Society. To a man, they all stated that given the same weather information that Captain Blanchard had been given, they would have done exactly the

same thing. Two steel-hulled steamers left both Boston and Portland the same afternoon. They were able to ride out the storm at sea, but both were deep-draft propeller steamers.

Myth: The *Portland* should have run for the safety of Gloucester Harbor.
Reality: When the full fury of the storm hit the *Portland*, she had already passed Gloucester Harbor hours before, and could not turn around to go back. Steerage also could not be kept by backing down away from the waves, as this would also turn the ship broadside to the waves.
The only thing to be done with what little effect the rudder had was to continually face into the waves to keep the damage to a minimum. To back down would entail stopping the engine and running it backwards to go in reverse. Walking beam engines had no transmissions as in later ships, and could not be reversed easily, especially in a huge sea.

Myth: The ship sank outside Peaked Hill Bars.
Reality: The hull of the *Portland* lies more than 24 miles to the north of Provincetown, off Cape Ann in more than 400 feet of water. The wreck found in 1945 by Snow's diver, Al George, in 144 feet of water and seven-and-a-half miles from Highland Light, turned out to be a mostly buried wreck of a five-masted schooner. That was the reason the diver was unable to wrap his arms around the mast. The *Portland's* masts were only about twelve inches in diameter.

Myth: The *Portland* collided with another vessel, either the *Pentagoet* or the granite schooner *Addie Snow* from Rockland, Maine.
Reality: During July and August of 2002, the wreck of the *Portland* was scanned and videoed upright and sitting on the mud on the bottom of Massachusetts Bay with its hull intact and undamaged by collision.

Myth: Nothing from the steamer *Pentagoet* was ever found on Cape Cod.
Reality: None of the bodies from the *Pentagoet* ever washed ashore, but some of its distinguishable red-painted trim was discovered in with the *Portland's* wreckage, along with some items from the granite schooner *Addie Snow*.

Myth: The main wreckage and hull of the *Portland* was not found until 1989.
Reality: In April of 1899, one of the last vessels to see the *Portland* afloat during the storm, the fishing vessel *Maud S.*, was dragging off of Gloucester in more than 400 feet of water and snagged some wreckage on the bottom and brought up fittings identified as coming from the *Portland*. This information was scoffed at by retired Navy Lt. Nicholas Halpine, who was in charge of the search for the ship at the bequest of The Boston Globe and Governor Roger Walcott. As all of the wreckage had piled up on the Cape, the debris pulled up in the nets was thought to have been lost overboard during the storm. Thus, the true location of the main hull of the ship became lost to history.

Myth: The *Portland* had been illegally equipped with the inferior Tulle lifejackets.
Reality: Tulle life-jackets and vests actually contained hollow reeds that floated and were tied together in short lengths to form life vests, not lifejackets, as are used today. The Tule reeds eventually get waterlogged and start to sink. The *Portland* had none of these inferior vests aboard and actually had more than 600 of the far better cork-filled life-jackets aboard, more than three to choose from for everyone aboard at the time. In the freezing seas, however, these only prolonged the agony.

Myth: A passenger list had been left ashore at the time of the sailing.
Reality: The only complete list of passengers sailed with the ship, so it is impossible today to know the exact number of passengers. 192 passengers and crew of the ship is the best guess having been arrived at by many years of research of the noted marine historian Edward Rowe Snow. A duplicate passenger list is now retained shore in case of the future loss of another passenger ship. So some good did come out of this loss.

Author's Image Collection

Building the Portland 1889 (close up).
Steamer *Portland* being built on launching ways 1889. Author's Collection

Building the Portland with other ships.
New England Shipbuilding Company Bath Maine 1889. Author's Collection

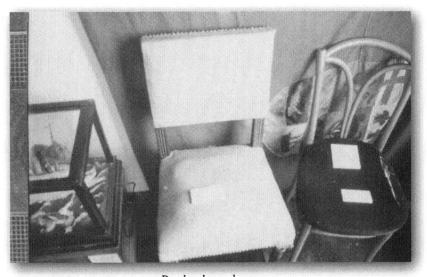

Portland wreckage.
Two chairs salvaged from Steamer *Portland's* upper decks found on beach in Truro, Mass. Author photo Truro Historical Society

Upper section of Steamer *Portland* hitting Peaked Hill Bars reef.
Just before breaking up. Artist Dave Moore. Author's Collection

Pilot Schooner Columbia wreck.
Ship smashed ashore at Sand Hills, Scituate, Mass. All 5 crew lost. Author's Collection.

Coast Guard Cutter *James*.
New 2015 named after famous Hull, Mass. lifesaver Joshua James. Author photo.

Imaging *Portland* wreck.
Remote operated vehicle Hela, being hoisted overboard to dive to Portland wreck 500 feet below. Author photo.

Author wreath laying.
Art Milmore holding wreath about to be laid over grave of Steamer Portland. On left is Mr. Ben-Cowie Haskell, expedition leader.

Wreath laying.
Author tossing wreath overboard from RV Connecticut over Portland. Science Channel crew at right. Dave Truby photo.

Artist Dave Moore.
Finishing drawing for author of Captain and Officers wrestling with wheel to keep on course. Author photo.

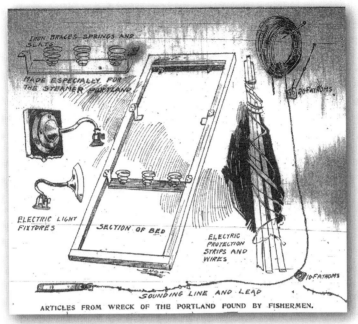

Portland wreckage hauled up in fishing nets.
Articles confirmed by Portland's former crew as being genuine. Courtesy Boston Globe 1899. Collections Boston Public Library.

Artist William Mueller at drawing board.
Painting ocean liner. Author photo.

Portland model.
Beautiful model of Portland. Courtesy of Centerville Historical Society, Cape Cod, Mass. Author photo.

Portland's main steering wheel.
Actual powered wheel from Portland salvaged from wheelhouse. Name added later. Courtesy of Centerville Historical Museum. Author photo.

Portland then and now.
Top 1890s, Boston Harbor. Bottom wreck today. All sections matchup.
Courtesy of Jerry Studds, Stellwagen Bank Marine Sanctuary.

Portland's Captain.
Captain Hollis Blanchard. Just promoted to Captain two
weeks prior to its loss. Author collection.

More *Portland* wreckage.
Captain Snow of Orleans holding dual spare wheels with other Portland wreckage. Courtesy of E.R. Snow archives, Mugar Library, Boston University.

Gallery Deck dishes.
Close up of Portland wreck. Courtesy of Dr. Ivar Babb, Director University of Connecticut. National Underwater Research Center.

Fighting the wheel.
Captain and Officers fighting the twin emergency wheels to stay on course in high seas. Dave Moore artist. Author's collection.

Portland Diving group.
Reached the wreck of Portland in 2008. L-R Don Morse, David Faye, Paul Blanchette, Slav Mlch and and Bob Foster. Bob Foster photo.

Evening shipping on Boston Bay 1891.
Pilot boat Columbia passing Steamer Portland. Atrist William Mueller. Courtesy Minnesota Maritme Museum Collections.

Joshua James and Crew.
Front row L-R Jim Murphy, Joshua James, Jim Dowd, George Pope. Rear L-R Francis Mitchell, Martin Quinn, Matt Hoar, John James. Historical N.E.

Cape Cod Morgue.
Temporary morgue, Mayo's Blacksmith Shop, Orleans, MA. Courtesy of Boston Public Library Collections.

Hull Lifesaving crew and boat.
Point Allerton Lifesaving Station. Courtesy Hull Lifesaving Museum Collections.

Portland lifejacket and light bulb.
Mr. Philip Smith of Weymouth Historical Society wearing Portland lifejacket.
Author photo. Courtesy Weymouth Historical Society Collections.

Portland at Dockside.
Showing rudder, paddle-wheels, and height of ship above water. Author's Collection.

Art Milmore

Portland in Boston Harbor.
Slowly coming into India Wharf. Author's Collection.

Portland at Dockside at India Wharf Boston.
35 foot high bow. No match for 40 foot waves. Courtesy
of Matthew Lawrence and Deborah Marx.

Portland wreck first time published.
Showing intact crank rod left from walking beam down to paddle shaft. Courtesy of Gary Kozak of Klein Sonar, NH.

Crew of Steamer *Baystate*.
Portland's Sister Ship. Captain William Snowman upper right, first Captain of Portland. Courtesy Virginia Snowman Family Collections.

Sara Fuller in Hawaii.
Looked for Whitten half of family for 35 years. All together now for first time in 100 years. Sara Fuller photo.

Steamer *Horatio Hall* off Provincetown.
In eye of cyclone, mistaken for Portland, survived storm. Author's Charles Williams Album.

Steamer *Kahtahdin* 1888.
Limping into Portsmouth Harbor, NH after gale. Coal ran out, had to burn cargo and parts of ship to survive. Author's Collection.

Jerry Studds Stellwagen Bank Marine Sanctuary Inside
dark lines dark area in water over 400 feet deep.
Courtesy of Jerry Studds Stellwagen Bank National Marine Sanctuary.

Portland wreck, stern at left, planking missing.
Engine boilers and boiler uptakes, stacks missing, bow on right. Arc shape in mud possibly missing starboard paddle-box. Courtesy Gary Kozak Klein Sonar, NH

CHAPTER 16

The *Portland* Victims – The Human Cost

THERE WERE NOT ENOUGH OFFICERS aboard the ship to man all the lifeboat stations in case of an emergency. This opinion was offered by retired former master of the *Portland*, Captain Craig, to reporters in Maine after the loss of the ship. If the ship sails could have been partly flown, they would have at least have steadied the ship from rolling so much and taken longer to have been damaged by the seas. Eventually, they would have been ripped to shreds by the over one hundred mile per hour wind gusts.

More than enough of a coal supply had been carried aboard for both legs of the voyage to and from Maine. Even if the fuel supply had run out, the cargo, furnishings, or even parts of the ship could be burned to stay headed into the seas. The *Portland* was designed to be the most seagoing of all the Maine built paddle wheelers. By having a much higher freeboard it encompassed the sponsons, extending out from the sides of the ship, fully enclosing them to protect them from damage in a high sea, and to add streamlining to the hull.

Paddle-wheelers did not have portholes as in a more oceangoing ship. They had instead, large square glass windows that were thin and could be easily smashed in by a large sea. Consequently, this introduced large amounts of water rolling across the main salon, creating havoc with the passengers. Sidewheelers were not only the large steamships lost in the storm. The 128-foot steam freighter, *Pentagoet,* also went down with all hands, and it was propeller driven. Even the best of captains cannot delay the inevitable sinking forever. Inside Boston Harbor, the more than 400-foot steel steamship *Ohio* was driven ashore on Spectacle Island, as her officers had barely 50 feet of visibility in the driving snow.

The best of the main steering wheels found on the Cape was bought and owned by Charles Lincoln Ayling of Centerville, near Hyannis. I found in the back of Edward Rowe Snow's 1943 book, "New England Storms and Shipwrecks" in his notes on his extensive chapter on the loss of the *Portland.* In the spring of 2003 I called the Centerville Historical Society, which reported that it indeed had the wheel on prominent display in its collection.

Upon his death, Ayling had donated the wheel and one third of his estate to the historical society to build a new wing on the building and to endow a fund for the society's upkeep and preservation of its many historic artifacts.

With the loss of all of the crew and passengers with the ship, the families of the dead were all thrown into disarray. Some families were ripped apart, but the African-American crew members families were worse off than the rest. This later led to the closing of the African-American Abyssinian church on munjoy Hill in 1914. Thankfully it survived to this day and is being restored. They were the hardest hit of all.

CHAPTER 17

The First Search for the *Portland*

HUNDREDS OF FEET BELOW THE surface of Massachusetts Bay and enclosed by the Stellwagen Bank Government Marine Sanctuary, the main hull of the *Portland* lies upright at peace on the muddy bottom. The hull appears to be mostly intact with little damage showing upon impact on its journey to the ocean's floor more than 100 years ago. Known as New England's "Titanic," the ship of mystery is finally giving up some of her long-held secrets.

Since no one who lived through the massive storm saw *Portland's* final moments, the mystery of the exact cause of the sinking has existed for more than a century. The latest side-scan images by the Stellwagen Bank Marine Sanctuary staff show the walking beam, or connecting beam, on top of the massive 2,000-horsepower engine. This engine was stopped in the extreme forward, or down position. In books on 19th century marine engineering, if the engine for any reason stops in either the top dead center or bottom dead center positions, the engine cannot be started in either of these positions without manual help. To restart the engine with the piston in either of these positions, a panel on the inside of the paddle box must be removed and a 4x4 timber must be inserted through the opening to lever the paddlewheel to a different position. Then, a steam intake valve can be opened to introduce steam into the cylinder to restart the engine.

If a crewmember attempted to remove the port cover in 40-foot seas, the incoming blast of icy seawater would blow him off his feet, and most likely the panel would not be able to be closed again due to the fierce deluge. The stopping of the engine in this position in 40-foot seas for any reason would be fatal due to the impossibility of recovering control fast enough to keep the seas from blowing the superstructure off the ship, resulting in immediate sinking from down flooding.

The only two plausible scenarios that would cause the engine to stop in this fatal position are seawater filling up the ship from numerous leaks in the bottom, or through damaged or missing pieces of the superstructure, causing the water to come in contact with the lower part of the cylinder. This would make it warp slightly by contracting on contact with the freezing water, and the piston would jam way down at the bottom of the cylinder and lock in place. Alternatively, if the eight-inch thick crank rod broke, causing the engine to run away (or over-rev in modern terms), the flailing crank rod would move in a much larger arc, destroying everything in its wildly swinging path.

Either scenario would be immediately fatal to the ship, with an almost instant loss of power to the pumps, steering, lights, etc. The ship, in immediate dire straits, would have less than 60 seconds to live. In more than 400 feet of water, it is too deep to anchor, even if one of the ship's iced-over anchors could be freed in time. The ice would have to be chopped off with a sledgehammer, and a crewmember would be endangering himself in the process.

In the future, exploring the wreck site would begin like this: A small submarine would come in along the mud line using strong lights and sonar. Months after it sank, the *Portland's* wreckage became entangled in the fishing nets of the schooner *Maud S.*, one of the last ships to sight her the night before she sank. Some of this debris was brought into Boston to be identified by officials of the Portland Steamship Company as having come from the wreck.

As most had been looking for the wreck off Cape Cod, it was thought to have fallen from the ship during the storm and was not from the actual wreck. This was only one of the facts that John Fish and Arnold Carr used in their high-tech search for the wreck location in the 1970s and '80s. Coming across the mud line, the bottom starts to look distressed by going from smooth mud to disturbed clumps. These are the result of the impact of a 280-foot ship weighing more than 2,000 tons impacting on the soft muddy sea floor of Massachusetts Bay. The submersible would now come upon a vertical wall about 20 feet high partly covered with fishing nets and cables. Any wreck site is usually teeming with fish and other marine life as it becomes an artificial reef.

The bow never opened up when the hull hit bottom damaging the hull from the main deck down. This is all that was left. The savage seas had ripped off the superstructure and all of the passenger accommodations above deck during the sinking. Proceeding toward the mid-section, at deck level, a person would next come upon the main cargo- loading hatch in the main deck. All of the freight-type cargo would be located and stored on the lower deck forward, as it would be the least comfortable part of the ship. The most comfortable part of any ship is from the midsection back to the stern. The crew's quarters would also tend to be in the bow area. The best parts were reserved for the paying passengers. Going further aft along the main deck, you would come across the uptakes for the boilers, topped by the stubs of the twin side-by-side smokestacks, which identify this wreck as the *Portland*. A few meters aft of this would be the 50-foot high wooden A- frame that encased the one-cylinder steam engine.

At the bottom of the double parallel A-frame would be the engine control panel, where the engineers stood when they ran the ship. At the top of the panel should be a shield with the number 57 on it, denoting the number of the engine installed early in 1890 by the Portland Engine Company. At the top of the A-frame, also entangled in nets, would the 22-foot-long walking beam that rocked back and forth constantly, transferring the up-and-down motion of the engine piston to the crank rod, which turned the main paddlewheels.

The holes where the steel paddle boxes used to be come into view next. The wooden paddle-wheels made of oak had long since rotted away. On super-close inspection with an ROV, it could be seen that the main shaft bearing made of lignum vitae, a super hard wood from South America, are either missing or took a terrible beating from the storm. Proceeding still aft to the stern area, most of the ship's decking is in this area should be mostly intact, including the deck bits used to tie up the ship and the chocks which the ship's lines were threaded through to reach the bits on the main pier.

At the extreme aft end of the ship still hangs the ship's rudder, turned all the way to one side in its last fatal fight to turn back into the wind. The courageous engineers and firemen in the bowels of the engineering spaces did their utmost duty to the end, even though they knew they could not get out in time when the ship foundered. This concludes the tour of the wreck of the Steamer *Portland*; the most palatial steamer ever to cruise the coast and rivers of Maine. We will never see her likes again, except in pictures. The last of the Maine steamboats ran until 1923, when trains and cars made it faster and cheaper to travel the 100-mile distance to Portland over much improved roads. This type of passenger luxury only exists on the modern liners of today.

CHAPTER 18

The New *Portland* Searches and Victim Stories

ONE OF THE SADDEST STORIES of the *Portland's* sinking concerned one of the families of those who went down with the ship. John Whitten, a watchman on the *Portland* from Milbridge, Maine, left behind a wife named Letitia, three sons and a daughter named Audrey. With the sudden loss of her husband, along with the rest of the passengers and crew of the *Portland*, she was left with no funds with which to feed her family. As life insurance did not exist in those days, she had to find work immediately. As she was finding it quite impossible to care for her family, Mrs. Whitten had to make the most difficult decision of her life. She had to put her daughter, Audrey, up for adoption. Dr. John Thompson, a physician from *Portland*, adopted her. Audrey's granddaughter Sara, who lives on one of the islands in Casco Bay near Portland, has been unable to find any of her relatives from the Whitten family.

Another tragedy befell the Portland Abyssinian Church on Munjoy Hill. Nineteen of its 20 crew members went down with the *Portland*, leaving behind wives and children to fend for themselves with the sudden loss of the family income. As two of the crewmen were also trustees of the church, this hit the congregation especially hard, ultimately resulting in the church's closing in 1914. In addition, the First Parish Church of Portland lost its leading soloist, Miss Emily Cobb, in the sinking. She was to have sung for the congregation that Sunday. What made it even worse was that like most, her body was never recovered after the wreck. Many relatives of the dead journeyed to Provincetown to look for the bodies of loved ones. Only 40 families were successful.

I believe that some of the heavily sanded-in bodies of the victims are still there today, buried deep under the sand on the beaches of the outer Cape. In the Hull, Mass., cemetery, there is a gravesite for unknown shipwreck victims called Strangers Corner. It holds the remains of more than 100 dead mariners, some from the *Portland* storm. Although some of the bodies were carried with the wind and waves floating on top of the water, no *Portland* lifejackets were found on them.

A month after the disaster, the Portland Steam Packet Company gave each of the families of the dead crew a month's pay to help them in their grief. This is all the compensation they received. After the *Portland's* loss, the line's slightly smaller spare boat, the *Tremont,* was put on the line opposite the *Bay State*.

Captain Alexander Dennison of the *Bay State* was now the most senior captain of the line until a new replacement captain for the Tremont could be found. Officials even considered bringing someone out of retirement. After much conversations in Maine steamboat circles, a replacement propeller-driven ship with much higher sides, the *Governor Dingley* was ordered built at a yard in Delaware. It would not be ready for at least a year. It was especially designed for high seas storm conditions. It would be later augmented by the addition of a two near sister-ships, the *Calvin Austin* and the *Governor Cobb*, which ran until the 1930s. The renowned New England Shipbuilding yard closed its doors in 1906 and was later replaced on the same site by the Texas Shipbuilding Company. The New England Shipbuilding name was later resurrected with a yard in South Portland, building Liberty Ships during WWII.

One of the victims who went down with the *Portland* was Merton L. Small of Woodfords, Maine, near Portland. Small, a 26-year-old single plumber had been visiting with family in Cambridge. He had decided to spend an extra day with his family and old friends. He was well known in the Boston area, having had lived in the vicinity and having had worked at Woolworth's. A married couple, Mr. and Mrs. Arthur Hersom, also was on board. Mrs. Hersom had come to Boston two weeks previously to attend to her ill mother, who lived on Arlington Street in Boston, and was later joined by her husband. They had intended to stay for Thanksgiving and return on Saturday. Mr. Herson was 29 years old and was a salesman for the Hammond Beef Co. in Portland. They left no children.

Albert Carter, age 22, one of the firemen on the ship, was a four-year veteran of the crew. He left a wife and two children. Mrs. Horace Pratt and her daughter, Amy, a promising musician, only 17, were also among the lost souls aboard the vessel. One of the many single men aboard was William Roach, of Portland, only 23.

Two sisters from Weymouth, Mrs. A.S. Chickering and Mrs. E. Augusta Wheeler, were on their way to Augusta to attend the funeral of a third sister, Ms. Mary Lara, after receiving a telegram announcing the death. They immediately traveled to Boston to make it aboard the evening steamer to Portland. From there, they were to continue to Augusta by train. Their bodies were later recovered on the outside of Cape Cod. Mrs. C.E. Harris, one of the African-American stewardesses on the vessel, belonged to the Abyssinian Church in Portland. Two other prominent citizens of Portland, John Murphy and Oren Hooper (and his nine-year-old son) were aboard, as were Miss Sophie Holmes, a young schoolteacher in Portland. She was very popular lady of refined tastes and a lovable character, according to her obituary in the newspaper. She was returning from Mercer, where she was born, to take up her school duties in Portland. Mr. F.A. Ingram, the 47-year-old purser, had worked on steamboats for 20 years. He left behind a wife and sister. His sister was the wife of the line manager, John Liscomb. Robert Foden, age 10, was accompanying his mother from Cincinnati to Portland, their former home. Both perished. John Albert Dillon, a 44-year-old oilier on the *Portland*, was born in Eastport, but lived many years in

Portland. He left a wife and two sons. Miss Madge Ingraham, who worked at M.M. Bailey's in Woodfords, took the ill-fated steamer on her return trip from visiting her home in the south. The list of passengers goes on and on.

A benefit concert arranged by Mayor Randall of Portland and chaired by the noted Professor W.R. Chapman, raised money for the victims' families. Subscription papers were also circulated to take donations from the public, as the generosity of Christmastime would be most heartfelt. The city of Portland really came together for the victims' families.

CHAPTER 19

Expedition to the Portland, September 13-18, 2003

ARRIVING IN GLOUCESTER DURING THE week of September 13, 2003 after its journey from its home base of Groton, Conn., the research vessel *Connecticut* tied up at the state pier in Gloucester's inner harbor. Operational checks of the navigational equipment and its computerized positioning systems were carried out at dockside before heading out to sea to the *Portland* wreck site more than two hours away.

On the first two days, searches were carried out using side-scan-sonar to image five shipwrecks that were to be explored during the coming week. Leading the search marine team was biologist Ivar Babb, head of the National Underwater Research Center, allied with the University of Connecticut. As I boarded the ship on September 14, I carried with me the album once owned by Charles Williams, the agent for the Portland Steamship Co. at the time of her sinking. A film crew from the Science Channel which is owned by the Discovery Channel was filming a special about the *Portland*.

I gave a 40-minute interview with the camera crew while waiting for the new ROV to be repaired. Our trip to sea was not to be that day due to electronic troubles with the brand new ROV (remote-operated vehicle). After working all day on the vehicle, engineers from the ROV Company got it back together and it was hoisted over the side for underwater testing in the harbor.

The lights, thrusters and all of the electronics finally tested out in first-class order, but it was too late in the day to head out to sea.

The next day, when I could not be aboard, the ship headed out to sea. The ROV was used on various lesser underwater wrecks as a tuneup for its try at the wreck of the *Portland*, scheduled for the next day. On the following day, the ROV spent many hours imaging the wreck of the *Portland*; for the first time on this wreck, a high-definition TV camera was used.

The same problems that they had faced the year before in 2002 were taken into consideration during the dive planning. As many fishing nets had covered the stern area of the *Portland*, the vessel came in from the bow of the ship at the opposite end.

In coming in over the bow, they noticed one of the ship's anchors sitting on the bottom near the bow with its chain running up onto the ship. In going back from the bow, a small anchor winch was filmed, and the forward stairway leading down into the crew-men's quarters could be seen. The ROV was physically too large to fit down the stairway to look inside. Further aft, they then came up to what looked like the ship's smokestacks, but on closer examination, they turned out to be the uptakes from the boilers. You could see from only three feet away where the two smaller diameter smokestacks had broken off.

The openings for the small smokestacks only seemed to be about 24 inches in diameter, but there was one for each boiler. In their 1890's appearance there was a rod between them to cut down on the vibration that would occur at top speeds. Further aft, the walking beam was visualized close up, but the ship's bell containing the name *Portland* on it could not be seen. The crank rod going down to the paddle shaft was intact at its top end.

Late in the dive, the light-bar on the ROV got caught under a pipe, trapping the vehicle on the wreck. All kinds of maneuvering were tried to free it, but to no avail. The only thing left to do was to haul up on the cable to free it, but this resulted in the breaking of the tubular aluminum light bar. The vehicle surfaced with the light-bar hanging off one side.

That night it was taken ashore to a welding shop and it was repaired almost as good as new. The next morning as soon as the bar was back aboard the ship the *Connecticut* left the dock for the two-hour journey to the dive site. This day, September 17, would be spent exploring the two coal schooners, the *Palmer* and the *Creary*, to see who hit whom in December of 1902, causing them both to sink, not five minutes apart, as reported in the newspapers of that time. They were still stuck together in an embrace of death.

Even 100 plus years later, there are the remains of three masts still partly standing, but are broken off about 20 feet above the deck. Much time was taken filming the lower part of the hulls, where the wood appeared to be soft and mushy. However, every time they raised the machine to the level of the main deck, they ran into the many fishing nets that completely cover both wrecks. Only at the stern of the two ships were the scientists able to visualize the main deck slightly. Further video was taken on the wreck of a fishing trawler that sank nearby. As the *Connecticut* headed back toward the shore at the end of the day, the ship stopped over the grave of the *Portland* to allow me to lay a wreath over the side-wheel steamer, which my mentor, Edward Rowe Snow, had wanted to do for years. Mr. Snow passed away in 1982 and was never able to lay the wreath himself due to the unknown location of the wreck.

At the beginning of the wreath-laying ceremony, I produced a green bay-leaf wreath with a banner across it containing the words "Steamer *Portland*." In my left hand was a short speech written on a crumpled napkin that I kept saying over and over again to myself so I would not mess up the words on the Science Channel's camera. At the moment the words finally sank into my memory as I uttered these words:

> *"In memory of a good friend and mentor, Mr. Edward Rowe Snow, who kept the Portland story alive, and to the 192 passengers and crew who died here, may this spot forever be a sanctuary for their souls."*

I then threw the small wreath overboard in front of the entire ship's company and watched it float gently away on the waves. I was somewhat disappointed as the wreath flipped over in mid-air and landed face down so the banner could not be read. Later, one member of the film crew said, "Even though we could not read it on the water, the victims down below could look up and read it." As the ship picked up speed and headed west for the two-hour journey back to shore, I thought to myself, "What a day!"

Although the exact cause of the sinking could not be found during this expedition, we were very close. In a future expedition to the still-secret *Portland* site, some of the netting on the stern of the ship will have to be cut away to reveal the lower part of the main crank rod going down the paddle shaft, then one of the final pieces of the 102-year-old *Portland* mystery will fall into place. We all have become undersea detectives, and I absolutely loved it.

In a two- day seminar held on Nov. 7 and 8, 2003 at the Southern Maine Technical College in South Portland, where I was an invited speaker, the video footage shot by the Science Channel during the September 2003 Gulf of Maine expedition was first shown to the public.

Coming along the mud line to the bow of the ship, the *Portland's* anchor could be seen hanging from the bow in the upright position leaning against the straight stem of the ship. The anchor chain ran from the top of the anchor up onto the main deck of the ship to the anchor winch. The clarity of the high-definition television was astounding. When the ROV reached the galley area, one could almost read the lettering on the bottom of the drinking cups scattered about. I questioned the previous year's diving expeditions' findings of the whereabouts of the *Portland's* smokestacks – though it looked on the side-scan sonar images that they had survived, when an exploding boiler blows off a ship's superstructure, the tall stacks are usually the first things to go.

Under very close examination, what looked to be smokestacks were found to be the uptakes, or the chimneys, for the boilers that were five feet in diameter. At the top of these, on the inside edges were round holes, 24 inches in diameter where the stacks themselves had been. It was nice to be finally vindicated.

The main deck on the starboard or right side of the ship, looking forward, is partially collapsed and there are holes though the main deck where the sea worms have eaten through. The main walking beam at the top of the engine is mostly covered with sea growth and the remains of the fishing nets. Most of the main deck over the stern area is missing. A paddle box, that American Underwater Search and Survey videoed in the late 1980s, is now missing. It was probably ripped off by fishing nets and lies somewhere within three quarters of a mile from the wreck. In the side-scan sonar images leading up to the wreck, some large objects on the bottom can be seen nearby, but these have not been looked at by video cameras.

After sitting on the bottom of Massachusetts Bay for 100 plus years, the wreck is rapidly deteriorating to the point where the shipworms will have eaten all of the wood and only the iron will remain. The expedition tried to disturb the wreck as little as possible, as it is a gravesite for the 192 passengers and crew, who will forever rest in peace.

CHAPTER 20

Side wheels verses Propellers

After the loss of the *Portland* with all hands on November 27, there was some argument and soul-searching about whether to build any more paddle-wheel boats. They operated well in the rivers, but left a lot to be desired in handling during an ocean storm. Paddle wheelers had been used to cross the Atlantic Ocean since the 1840s with no losses attributed to their design. They had very high sides, with a minimum of overhang, and the paddle guards did not stretch the whole length of the hull, like the coastal steamers. The steam engines at the time became ever the more reliable as better metallurgy was used to manufacture the parts. By the time of the *Portland's* construction, the cast iron parts were almost flawless.

Four steamers were out at sea in Massachusetts Bay on the night of November 26, 1898 – the *Portland*, the *Horatio Hall*, *Pentagoet* and *Gloucester*. Two survived, and two did not.

The *Portland*, which was a side-wheeler, and the *Pentagoet* a propeller-driven freighter, were both lost. This would seem to cancel out each argument about which ship was more dependable.

Here is the account of Captain Albert Bragg of the *Horatio Hall*, as told to a reporter from the Portland Press Herald on December 3, 1898:

"The handsome steamer Horatio Hall of the Maine Steamship Company, which left Portland Saturday evening for New York, about the same time *that the ill-fated steamer Portland left Boston for Portland and went through the hurricane of Saturday night and Sunday without any damage at all, arrived here (in Portland) yesterday looking as spick and span as she did the day she started out on her trials. There was not so much as a deck bucket missing from this handsome ship, and though she went through the hurricane which wrought such havoc all along the coast, and was for hours right in the storm center.*

This shows better than anything else what a great steamboat the Hall is. She has proved herself to be one of the safest steamers along the coast, and she was also one of the best handled. It was about 3:00 p.m. when the Horatio Hall tied up to the dock here (in Portland) on her return from New York.

We left Portland about 7:56 Saturday night, and went outside where we found the sea smooth and a moderate breeze was blowing. At this time, while the weather looked a little threatening, we had no idea what was in store for us before morning. We kept on our course toward Cape Cod and about 10:30 p.m., it began to breeze up a little from the east-northeast. When we were a little way out past Boon Island, it began to snow. This being about 11 p.m., we continued on our course south by west for the Cape and continued to run at full speed. At midnight it was blowing heavy from east-northeast.

The sea began to come up fast and in a little while we were running through a tremendous sea and shipped some water forward. The wind continued to increase in fury and I knew we were in for a very bad night of it. The barometer was constantly falling. At 1:30 a.m. Sunday morning we were 30 miles from Cape Cod and I told Captain Harding, the pilot, that I believed we had better heave-to (pushing the bow into the wind).

We brought the nose of the ship up into the wind without any trouble and heaved-to at this time. Cape Cod lay in a southwestern direction at this time. The wind continued to increase in strength and at 2 a.m. it was blowing a hurricane, with the barometer dropping off in a startling manner. The Hall behaved splendidly. I never knew of a vessel which stood the heavy seas as well as she did. Of course, we shipped a little water but not any great amount of it at a time. I knew we were making a southerly drift because the seas were so heavy that we could not get the Hall right up into the gale. The sea was terrible.

I have been going to sea for 30-odd years, but I never saw the waves run as high as they did on Sunday morning at about 3 a.m. It did not seem possible that these enormous seas could be kicked up in such a short time, even with a hurricane blowing as strong as it was. The barometer dropped down until it was 29.08 at 6 a.m. Sunday morning. The hurricane was now at its worst. At this time, I told Captain Harding, the pilot, which I did not think the wind would not hold up as furiously as it was then blowing for very long, and I was right, for at 7 a.m. the barometer went up a tenth.

But, the seas began to rise higher and higher. They went fully 10 feet above our flagstaff every time. I never in all my life ever saw such waves as we were then riding over and we didn't ship very much water, either.

At 8 a.m. Sunday morning, the wind suddenly died away and for some time there was little, if any, wind to speak of. At half-past eight Sunday morning we got the ship back on her course and headed southwest. In about an hour we made Nauset and Chatham right ahead. We had drifted, in seven hours, 40 miles to the southward, with our engines going hard enough to keep our head up into the wind. In all of this night, we did not cite a single vessel, nor hear the sound of one.

At 7 a.m. Sunday morning, we could not have been very far from the steamer Portland, which must have drifted down from this point before she foundered, but we saw nothing of her. When we ran in toward Nauset, we made out a fisherman running along with a bit of her foresail showing, but she seemed to be all right.

At 9 p.m., we hauled offshore for Pollock Rip Lightship. Here we have to go through a slue not more than 1500 feet wide. As we approached it, the waves were breaking on each side of us, and in fact, in front of us, five-and-a-half fathoms of water (33 feet). It had cleared up for a time and stopped snowing, and this was a godsend for us, for I know unless we got over the shoals before it started to snow again, we might have to ride out another night at sea.

When we passed the Pollock Rip Lightship, she was in the right position, but the waves were breaking over her and she was having a pretty rough time of it. We pushed on at full speed and at Shovelful Shoals Lightship, it began to blow great guns again from the north-northeast, and began snowing again.

There was thunder and lightning, too, and you have no idea how the wind blew or how thick it was. When we got to the Handkerchief Shoal Lightship, we found that she had broken adrift and wandered six miles out of position. Pollock Rip Lightship must have drifted from its position after we had passed them. We went on toward the Cross Rip Lightship and found her in the right place, but she was putting her bow down level with the water with every wave. At 2 p.m. on Sunday afternoon, we anchored off Falmouth Basin, Vineyard Sound. It continued to blow great guns and until 1 a.m. Monday morning when the weather moderated a little. We got under way and arrived in New York all right and without any damage or injury to the ship. We were reported lost at one time, so I understand, but we were never in danger for a minute, though for a ship like the Portland it must have been impossible to live through it."

Captain Bragg was very much distressed over the loss of the steamer *Portland*, but did not care to discuss the matter with the reporter. He knew Captain Blanchard well and thought a great deal of him as a seaman and a brave man. He said the *Horatio Hall* must have been very near the spot where the *Portland* went down, but he neither heard, nor saw, anything of her, or any other steamship. Captain Bragg was enthusiastic in his praise of the behavior of the *Hall* and again told the press reporter that she was the "finest vessel he had ever been to sea in."

On the return trip from New York, the *Hall* ran along the Cape by Chatham, Nauset, and Race Point with all hands on lookout for bodies or wreckage from the *Portland*, but not a thing was seen. Captain Bragg said that in his opinion, the *Portland* labored so hard in the heavy sea that she sprang a leak and sank some miles off Race Point.

The *Horatio Hall* had 20 passengers aboard for this trip to New York. The passengers suffered a little from seasickness, but at no time were frightened or panic-stricken, as the Hall was so ably managed and rode out the hurricane so well. The *Horatio Hall* was later sunk in a collision at sea in 1909.

In the winter of 1869, the steamer *Cambridge* had been caught in a fierce gale off the Maine coast. The low-powered side-wheeler battled the storm for more than 12 hours and came through it successfully. Later, Captain Pierce, the first pilot on the steamer *Katahdin,* going from Bangor to Boston, found himself in the middle of a storm off Cape Porpoise, Maine, for 12 hours. In keeping her head into the sea, she had used up all her main and reserve coal supply and had to burn the ship's cargo, consisting of spool wood, which burned nicely. Next, furniture and non-essential parts of the ship itself were burned in the battle for survival. Finally, the wind changed direction and the seas subsided somewhat, and she was able to enter Portsmouth (N.H.) Harbor for respite and a new coal supply. She later continued to Boston, and was taken out of service for a short time to be repaired. She sailed until 1891, when she was towed to Nut Island in Quincy, Mass. for scrapping. There, the tough old ship was slid onto the beach alongside an old running mate, the *Forest City,* which was burned to salvage the metal parts of the ship – a sad ending for such a fine vessel.

Although paddle wheelers would continue to be built for some years afterward, they would be used mostly in the rivers and harbors along the coast, but were gradually phased out. After the Loss of the *Portland*, the new steel screw steamer *Governor Dingley* was ordered as a replacement. She would later be joined by two more near sisters, the *Calvin Austin* and the *Governor Cobb*. All of these ships, including the *Portland's* newer, near-sister ship the *Bay State,* would eventually become part of the Eastern Steamship Line. In 1910, the *Bay State* was rebuilt with higher hurricane decks and the main paddle shaft was lowered and the newer, more efficient feathering paddle-wheels were substituted. They no longer needed the huge paddle boxes that the older-style radial paddlewheels needed. This opened up more room for cabins for paying passengers. The steamers went from single to twin engines – these could be reversed for docking purposes and added to the safety by providing a backup engine.

In the early 1930s, trains and roads had taken away much of the steamships' business, and they gradually ceased operations and were laid up. The ships in good condition were kept in storage, and later were called back into service during World War II. Some were sunk in action by the dreaded German U-Boats. One ship from one of the Chesapeake Bay Lines, the *President Warfield*, survived the war, was sent to Europe. Renamed the *Exodus*, it ferried Jewish Holocaust survivors to the Middle East, in Palestine, that later became the nation of Israel.

The careers of these boats were long and distinguished, which is why the steamers were called the old reliables.

CHAPTER 21

Laying the Blame

—

AFTER ANY HUGE DISASTER, THERE are many recriminations and blame to go around. In the case of the loss of the steamer Portland, there were many to go around. Captain Hollis Blanchard, was known as the storm racer. On previous voyages, he calculated when to leave port in timing the arrival of an upcoming storm, with the length of the voyage leaving about three hours leeway to arrive in port before the storm would be forecast to hit. This was done with other lines also, but the Captain could return to port if he felt that conditions had changed too rapidly to continue. He could also duck into the nearest port for protection from the rapidly advancing storm.

It was far better to arrive late than not at all. Many other Captains did the same thing, but some type of schedule needed to be kept. Paddle-wheel ships had crossed the Atlantic Ocean since the 1840's with hardly any trouble except when using too much coal during bad weather. This had happened to the Collins liner Atlantic, which had to put into Halifax, Nova Scotia to re-coal while on the way to New York, on occasion. Weather reporting at the time was somewhat fragmentary, resulting in incomplete forecasts, which raised hell with the planning of the Portland's voyage to Portland Maine. A front could rapidly accelerate the speed of advance of an oncoming storm. This is exactly what happened when the combined storms from the Great Lakes and the Gulf of Mexico travelled up the U.S east-coast toward New England. It is why even the Horatio Hall was caught by the rapidly advancing gale while off Cape Ann. This fact caught almost everyone by surprise.

Only the skippers that did not go out at all, weren't. The winds in the storm travelled in veins, which blew down rows of trees and barely touched others. These veins of wind concentrated the super-structure's debris, instead of blowing it all over the ocean.. The Atlantic Ocean is a far more dangerous place than most people realise. The author having crossed it twice on a U.S. Navy destroyer, has seen it all too well, especially coming back across in December. Every three hours or so it looks completely different, on some days. Everything in the Portland Steamship Line at the time was in disarray, with everyone either in a new post or away from their jobs at the most critical time. So there was a lot of blame to go around. Some things can be blamed on an act of god, but acts of man can change things for the worse. .So many bizarre things happened during this storm, nobody could even figure it out. Mr. Jot Small from Cape Cod said that so many odd things happened in the thirty-six hour gale,.that it will always be a mystery, and will never be solved. She is sunk out in deep water, far from the Cape, and that she sank like a sounding lead. His conclusion was the closest to the truth. Even in modern times, ships still disappear without trace.. This is the lore of the sea, as well as the reality. No mode of travel can be made completely safe, there will always be a degree of risk. One can only learn from the past by studying history, so events will not happen that way again.

CHAPTER 22

The Sea Takes No Prisoners

IN ADDITION TO THE STEAMER *Katahdin,* which survived a ferocious storm off the Maine coast, were two additional steamers, the *Cambridge* and the *John Brooks*, which, both being side-wheelers, successfully fought the seas offshore and made it back to port.

The *Cambridge's* battle is detailed in an article published by the Portland Argus on December 11, 1898. According to Captain Johnson, "the wind came up suddenly. That explains the sailing of the steamer *Cambridge* of the Boston and Bangor Line. The *Cambridge,* being an old-fashioned side-wheeler, commanded by Captain Johnson, whose conduct during the trying scene of a terrible night won him well-deserved praise. The steamer made her way down the Penobscot River with no incident of unusual nature marking the trip. Following her stop at Rockland, she put out about 6 p.m. in the evening, passing Monhegan Island at approximately 7:30 p.m. The weather was somewhat rainy, but with no signs of anything unusual.

Shortly after the *Cambridge* passed Monhegan, the hurricane burst with all its fury, and it was determined at once by the captain that his only hope for safety lay in putting about (reversing course) and making for a harbor. This plan, however, was abandoned almost as soon as adopted, for the night was dark and the sea so tempestuous, that to approach the coast meant almost certain destruction on the rocks. The steamer was headed into the teeth of the gale in an endeavor to work offshore. The first accident occurred. The rudder braces gave way and it was quite impossible to keep the boat under control. The skill of the seamen was called into play, and an attempt was made to rig a drag, or sea anchor, so that the vessel might in that way be kept head to the sea. The attempt failed and the vessel struggled along, the rudder being used to steer the ship as well as could be done in its crippled condition. Minutes dragged like hours to the anxious passengers. At 9:30 p.m., they heard the loud hiss of escaping steam as the blow-off valves vented the tremendous pressure built up in the engines. Captain Johnson, cool through it all, informed those who asked him for hope, that the vessel and those on board were in the hands of a kind Providence in whom they must place their trust. Hour after hour passed, with the vessel wallowing in the great seas and darkness, and the passengers and crew desperately prayed for reprieve.

Finally, at 3:30 a.m. Thursday morning, soundings were taken and both anchors were dropped. Ninety-five fathoms of chain (552 feet) went rattling and smoking out through the hawse-pipe. Would the ship come to a stop? The ship's head wrung around, rising and falling on the mighty waves. The captain and pilots anxiously peered over the side, watching for the moment they could determine if the anchors held. The news that the hooks had taken firm hold on the bottom, the wind and sea no longer holding the *Cambridge* at their mercy, spread through the ship like an electric shock.

The place where the *Cambridge* came to anchor was near Little Egg Rock. There were anxious hearts on board when morning dawned and showed no more than half a mile astern of the steamer, the line of white marking the breakers. After a few minutes more of wandering, the steamer would have been thrown into the breakers.

Once there, her fate might have overtaken the *Portland*; only the bodies and wreckage being left to tell the tale of a night of terror off the Maine coast.

There was no change in the situation until noon, when the steamer *New England* appeared on the scene, coming around Pemaquid Point. She took the *Cambridge* in tow, and although three large hawsers were snapped like twine in the attempt, succeeded in getting the disabled steamer safely into Rockland. The only serious accident to those on board was received by engineer Hawthorn, who, while looking for a break in the machinery, fell and dislocated his shoulder. He suffered in silence until morning, when one of the passengers put the injured member back in joint. Captain Johnson, his officers, and crew received great commendation for the bravery and coolness they displayed. Through all the horrors of the night, the passengers, and especially the women, displayed much courage.

The steamer *John Brooks*, a side-wheeler long familiar to the travelers on the Boston and Portland Line, was also triumphant in her three-day battle with a fierce storm. The late Captain Liscomb (not the Line's general manager), commanded the steamer when she was caught out in a gale. It being prudent not to try to make port in the howling storm, Captain Liscomb headed the ship offshore and safely held her their hour after hour. During the time she was out, the vessel drifted from Monhegan Island down to the entrance of Massachusetts Bay.

These harrowing accounts indicate that the side-wheelers could take a terrific amount of pounding and still come through successfully. Anything can be sunk in such circumstances.

After the loss of the *Portland*, the opinions of many steamer captains were sought to ascertain what could cause Captain Blanchard to have put to sea as he did that fateful night of November 26, 1898. On April 9, 1899, the Portland Argus interviewed Captain Jason Collins of the steamer *Kennebec*. Captain Collins was well known and well respected, and on the same night he maneuvered the *Kennebec* from Boston Harbor, but was forced to turn back. Later that evening, he saw the *Portland* going out and heard her bells for the last time.

Of the controversy over the sidewheel steamers, and the assertion by many that this class of steamer should be driven from the ocean, Captain Collins thought this to be absurd. He failed to see the advantages of propeller steamers over the sidewheel. He said that of the many disasters during the previous few months, only the *Portland* was a side-wheeler, and more than a dozen propellers were overdue at that time.

Looking back to 1836, and it is a far cry from 1836 to 1898, he said that on this coast, the *Portland* was the only side-wheeler lost as a result of stress of weather.

> *"I have no doubt," Captain Collins said, "that Captain Blanchard would have taken the Portland into port all right; that she would have rode out the storm without difficulty had it not been for the fact that the big coal schooner King Philip, with 1,800 tons and perhaps more, failed to make Portland Harbor just before the storm set in.*
>
> *Had the King Philip succeeded in reaching Portland, in my judgment the Portland would also have gone in after the storm subsided. Captain Blanchard had a good boat and there would have been no difficulty in riding out a storm, assuming he encountered nothing worse than the storm itself. The King Philip was off the coast and was driven against the Portland.*
>
> *It is quite possible that the collision occurred off Peaked Hill Bars, and yet the hull of the Portland was located by the captain of the fishing schooner Maud S.*
>
> *"The Portland might have been struck and crippled, but not at once sent to the bottom. All the wreckage found after the storm might have been taken off then and yet the steamer was left leaking and crippled to drift to the spot where was believed to have been recently located.*

She would in that case have drifted on her natural routes. In going out as he did, Captain Blanchard acted on his best judgment. He made a mistake as thousands of the others have made similar mistakes. In my opinion, had he not met the King Philip, it would have not been a mistake.

There was nothing about the storm up until the time he sailed to have prevented his using his best judgment in going out. There was nothing in the weather report he had received up to that time to have kept him in port. Although captains deem weather reports to be valuable assistance in forming an opinion about the state of the weather, they do rely solely on official information.

Many times when a storm is coming rather slowly and we are sure to keep ahead of it, he paid no attention to the fact that it is coming to go out. Of course, we should not go directly into the storm if we could help it. The fact, as in other respects, the captain must form his own opinion and not on his own best judgment, possibly assisted by that of his pilots. I sometimes consult my pilots and sometimes do not. Every captain must be the final judge of what course of action to take. It is perfectly clear to my mind that in going out, Captain Blanchard had no idea more than any other captain of the real nature of the storm. He did not intend to take any needless risk. Of that I have not a particle of doubt.

There was not much sea when Captain Blanchard went out, and he had a short, while I had long run. It so happened that night that I had a considerable amount of passengers and I feared that they might find the night a bad one, not being accustomed to the sea in rough weather. But I went down to the lower harbor. I asked my pilots what they thought of it. We were then where we could see what kind of night it was and where we could turn back. One of my pilots said the trip will be very well, and the other thought it wasn't a very good night and we went back to our wharf. I remember that when I saw the Portland going out and heard her quarter bells, I thought that Captain Blanchard might run back, and yet they certainly did not think anything strange when he went ahead.

After the storm set in and we saw what it was to be like, we talked the matter over and decided that if it began to snow as early outside as it did in port; that it would be impossible for Captain Blanchard to make any headway. This matter of going out or staying in port when a storm is coming on or is reported is largely a matter of circumstance. I have gone out when the captains of the Portland Steamship Company have stayed in port, and again, they have gone out when I have stayed in. It all depends on the view each captain takes of the storm and of his circumstance.

I had the longer, and Captain Blanchard the shorter, run. I did not say what I would have done had I been in his place. I might have turned back, just as I did in my own case. I do not believe that in going out, Captain Blanchard acted otherwise than in accordance with his best judgment. He had a short run. There was absolutely nothing calculated to cause the loss of the Portland, except by collision.

Captain Blanchard was an experienced captain and I have no doubt his steamer was well handled. The report indicates that she was, and that she had actually ridden out the worst of the storm when she disappeared. Nothing occurred after that time due to stress of weather to have caused the loss. There is every evidence that she did ride out the storm just as we felt sure she would.

Having carefully considered all the facts bearing on the case as far as I have been able to obtain, I am therefore of the opinion that the Portland did ride out the storm. The loss was not due to stress of weather, that she was a strong boat, but that the disaster was due to the collision with the coal schooner King Philip (also lost in the storm with all hands). I am unable to see that Captain Blanchard was not fully justified by the *circumstances, the weather reports, the sea, and all he had to go by when he decided to go out. There can be no doubt that he believed he should be able to keep ahead of the storm, if indeed there was to be a storm, and that of any great magnitude, and that he fully expected to make a quick run to Portland."*

The long experience of Captain Collins gives his opinion all the more weight and authority. It shows that Captain Blanchard, in going out that night, acted on his best judgment and that not one of the other captains regarded the fact that he went out as anything at all remarkable. Captain Collins kept in port largely in consideration for the passengers who are unused to traveling by sea during a possible storm. That Captain Blanchard, as Captain Collins stated, had only a short run to make and was therefore not detained by the very consideration that carried great weight with Captain Collins.

It is noteworthy that the experienced captains and pilots in port that night did not for a moment believe that the *Portland* was in any real danger. They knew that Captain Blanchard could not make any port that night, but they did not fear that the *Portland* could go down.

During these years, deadly rogue waves, which can come up very suddenly and reach a height of more than 100 feet, were not considered. They continue to cause many ships to disappear without a trace, except for small bits of wreckage found afloat.

The person in charge of finding the wreck of the *Portland*, Lieutenant Nicholas Halpine, had publicly damned Captain Blanchard in one of the newspaper accounts following the tragic loss. Captain Blanchard's actions were vindicated by many of the steamboat captains along the coast.

When wreckage from the *Portland* was dragged up in the nets of the fishing schooner *Maud S.* in March 1899, it was shown to officials and crewmembers of the Portland Steam Packet Line. A crewmember identified one of the bed springs as being the same he recently repaired on the *Portland*. He knew his own unique work when he saw it.

The find occurred on Stellwagen Bank, in the northern section. Although wreckage is found in a certain location, it does not conclusively prove that it is the exact location of the wreck. Pieces can be lost and carried away before the vessel sinks. The wreckage was later found far north of the Cape, a distance of more than 20 miles.

As soon as word of the discovery hit the news, Lieutenant Halpine publicly labeled the find "rubbish" – in effect, stating that if he could not find the wreck, nobody could.

As Captain William Thomas Harpswell wrote to the Dearing Maine News and Enterprise in April 1899: *"Lieutenant Halpine says he can locate the Portland in two hours or in sixteen hours at the most, unless he locates her on the northern end of Stellwagen Bank in sixty-five fathoms of water (390 feet), fourteen-and-one-half miles from Eastern Point Light, Gloucester. It is a discredit and a shame upon the memory of Captain Blanchard to suppose that he would run his ship in such a blizzard as it was on the night of November 26th on a lee shore, knowing that there was not one chance in ten thousand of either him or any of his passengers getting off the ship. I knew Captain Blanchard well (Lieutenant Halpine didn't) and he was an experienced and capable sailor. He held his boat head on to the wind with a drag or sea anchor, working his steam as best he could until the steamer sunk, either by collision or from force of the storm.*

"She now lies on Stellwagen Bank. It is a wonder to me that any of the bodies touched Cape Cod. The bodies and wreckage drifted with the wind. The bodies started to drift to the head of the Barnstable (Cape Cod) Bay, taking a curved course. As the wind shifted to the north, they would drift out by the south channel. I have no doubt that many bodies passed within fifty feet of Cape Cod shore, but they never came up on the beach. If it had been clear, they would have been seen

and recovered. The fact that only forty bodies were recovered convinces me that many of them are in the wreck.

"Much of the superstructure is still attached to the hull of the steamer, or else it would have floated. She is so situated, however, that it will be impossible for divers to recover anything. Even today a depth of over three hundred feet is almost lethal, and on many occasions is. The place where we caught the relics from the Portland in our trawls is just where the wind and the sea would have carried her from the place I saw her at 9:30 p.m. on the night of November 26th. Lieutenant Halpine says that there was no ebb tide in Boston Bay that night, and he is right. But the force of the wind and sea carried far above the high-water mark.

As quick as the wind shifted to the north, this extra influx of tide had to come back faster and with more force than that which drove it in. This was responsible for the wreckage that was found along the outside of Cape Cod.

"No seafaring man of Captain Blanchard's experience would have tried to do what Lieutenant Halpine says he believes the captain of the Portland attempted. No seafaring man would have run his steamer before the wind that night, trying to make a harbor, because he would have had all he could do to keep his boat on top of the water, hove to."

This letter originally was written to The Boston Globe and was later reprinted in the Maine papers. None of the previous captains of the Portland Steam Packet Company had ever tried to run their steamers downwind in a storm. To a man, they all pointed the bow into the wind as the only way to ride out a storm safely. Trying something as dangerous as running downwind in a storm, even successfully, could result in a captain being fired by his line and losing his captain's license.

Another captain with great experience, Captain William T. Bacon of Towanda, Pennsylvania, wrote to the Maine papers about his thoughts on the fate of the *Portland*: "*The absence of all boats or parts of boats is an evidence to my mind that the Portland was sunk suddenly and quickly, as by collision and not slowly as though wrung, twisted, or broken by the force of the crashing waves.*

"*If the wreck of the Portland is ever located, I believe it will be well out in Boston Bay with some other unfortunate craft beside her. Who shall tell if the sinking was not brought about by the attempt to rescue some imperiled crew, the breaking of a connecting rod, the snapping of a wheel chain, brought under some sudden and terrible strain may have been the cause?* "*Let the cause have been whatever it may, we who know them for good and true men will always believe the officers and crew did their duty to the last, and that an error in judgment sailing would have pleased every passenger and their friends had the ship gone safely through, was no lack of knowledge or seamanship, nor wish to endanger the lives and property entrusted to their care.*"

CHAPTER 23

Solving The One-Hundred-Year-Old Mystery

―

When someone tries to solve a mystery that has existed for over one hundred years, the task is both daunting and personal. The biggest mystery in all of Mr. Edward Rowe Snow's books about the sea was also the most difficult. Mr. Snow, my mentor, worked on this mystery for over thirty-five years. However, the technology of his day was not up to the near impossible task ahead of him. In 1978, four years prior to his death in 1982, we talked at length about his lifetime quest to find the complete story of the *Portland*. By that time, many of the people he had interviewed that were out in the storm that night, had died. The information trail was drying up. He solemnly said "maybe in the future some new technology will be invented to help solve the mystery. He asked me that if I ever found the remainder of the material to complete the Portland saga, would I please do so". I gave Mr. Snow my word that this would be done. He was such a great teacher and mentor who showed extreme patience with me. By this time, Mr. Snow was becoming worn-out. He had lived one hundred and forty years' worth in his seventy-nine-year lifetime. I undertook a further thirty-two years of research on my own, finding more information that had been buried over the one hundred years. The Lord seems to put the right things in your path at exactly the time that you are mature enough and ready to accept them.

A lot of the facts in the stories that were written at the time of the sinking were accepted as totally true. In cross checking interviews and stories, a reporter wrote something that was incorrect and it was taken as the gospel truth. The weather service forecast the approaching storm from the Great Lakes, however completely missed the storm from the Gulf of Mexico, once it passed offshore. After that, the smaller storm ceased to exist. At this time, no thought was given to interview incoming ship captains about the storm offshore to possibly track it. Sometimes, half a weather forecast is worse than none at all.

According to a schooner captain off Provincetown, he had seen the *Portland* out to sea for two hours while the eye of the fast moving storm paused over Provincetown. According to the National Weather Service office in Taunton, MA, it is actually about twenty minutes. The erroneous two-hour information was believed by everyone, including the courts. Thanks to the genius of doctor Harold Edgerton of M.I.T., and his development of side scan sonar, shipwrecks on the bottom could be located much more easily and be seen in good detail. As time went on and improvements were made, the detail became fantastic.

The images of the wreck of the *Portland* on the northern part of the Stellwagen Bank National Marine Sanctuary, was the work of the superb Kline Sonar Labs of New Hampshire. The unit which resembles a small torpedo with fins on the rear edge is towed underwater near the location to pinpoint some identifying detail. An ROV-remote operated vehicle- was then lowered down to the wreck for live video. This usually results in either identifying the wreck as a certain ship according to its historical characteristics, to rule it out as a different wreck. In the case of the latter, the search begins anew. In the earlier searching of the area with a magnetometer, all the big hits were recorded; then the search shifted to one of the of the other hits.

On one of my two oceanographic trips to the wreck, I got to see ten hours of live video that no one had ever seen. We were seeing things for the first time ever. Even then, we did not see all of the wreck, as some parts of the ship are very dangerous to image. The ROV had been snagged four times already.

The loss of an ROV costing four hundred thousand dollars would immediately end the expedition until a replacement could be found. Fishing nets have been entangled on the wreck for the last one hundred years. There are also cables hanging above the wreck held up by the big glass floats. In case of entanglement, unable to go back-up or steer the ROV out of its dilemma, the pilots in the control van using brute force, attempt to wrench it free with the winch line, to try to free the ROV. Luckily, the somewhat rotted fishing gear gave way and up came the HELA, the ROV in damaged condition.

The wonderful crew of the research vessel *Connecticut*, under the awesome leadership of Dr. Ivar Babb and his crew can repair almost anything when the aluminum bar holding the lights were broken during the entanglements. The broken light-bar was offloaded at the end of the day, taken to the local welding shop in Gloucester, who repaired the bar overnight, and was put back and reinstalled on the ROV as the ship headed out to the wreck early next morning. I saw this with my own eyes as they put it back together. Over a five-day period, operating out of Gloucester, other wrecks would also be investigated to learn more about them.

In the summer of 2004, was left the dock at six-thirty in the morning and did not return until nine thirty at night. My drive up from Weymouth, on the south shore of Boston began at four A.M. I did not return home until eleven thirty that night, an hour and a half drive each way. I now began to understand what Mr. Snow went through in his research! Wow! We both had the same bull dog determination to see things through. Thanks to his years of work, much of the story is now known. Some of the answers will always be a mystery, as to why Captain Blanchard sailed at precisely seven P.M. into eternity.

Did a rogue wave suddenly turn the *Portland* sideways to be smashed by the seas? No other vessel saw her go down with all hands. Was she standing by another damaged ship to lend assistance? This could possibly be a reason. If some of the lifeboats had been successfully launched, they could not be recovered in the huge waves. The lifeboats, made of metal, with float chambers, could have survived in some form. Maybe, when they washed up on the African coast, the name *Portland*, may have been worn off by the elements, thus nobody could tell where the *Portland* came from. The possible wreck site, being in an area southeast of Gloucester, only reveals so much. There appears to be no evidence of a collision with another ship, as the main hull of the *Portland* appears intact.

There is no other wreck beside her in a southerly pointing direction, although there are some nearby. However, these wrecks are of more recent vintage and not from the time of the Portland Gale. The upper works of the ship were mostly ripped off, as the wreck down flooded from a combination of escaping steam from broken steam pipes and the down flooding seawater. Collapsing bulkheads and falling debris would have trapped many of the crew and passengers.

The massive weight of the engines and boilers were the final nail in the coffin, which caused the *Portland* to sink. The zero-degree temperature of the air as well as the twenty-degree temperature of the ocean water would kill any surviving passengers and crew in the water in a very short time. The veins of wind concentrated the debris and combined effect of the now northern direction of the wind, as well as the prevailing currents, drove the remaining part of the superstructure with its human remains southward on the outskirts of Stellwagen Bank toward Cape Cod.

When the superstructure hit the Peaked Hill Bar Reef off Truro, the structure which was breaking up, was found piled high upon a half mile stretch of beach. Most of the bodies of the passengers and crew ended up here with only some wearing their life jackets. Other bodies washed ashore further down the outer beaches, along with the pilot house and the huge 6-foot emergency steering wheels that would have been used in case of loss of power to the smaller main wheel. The smaller wheel, would have been used ninety-eight percent of the time. The six-foot emergency steering wheels were found lashed together. This led to the speculation that the crew had given up, however, that was corrected by the lifesavers and Portland line.

Some of the bodies that had been buried in the sand were later exposed during another smaller storm in early December of 1898. In total, only forty bodies were ever found. The remaining bodies were swept out to sea or entombed in the wreck of the *Portland*. The Portland line asked relatives to send photos of their missing kin so that the bodies could be identified.

In interviewing the survivors of the storm of November 26 and 27, 1898, many years later, their memories are somewhat altered by the passage of time. This can cause some confusion into the facts of the story. If artifacts are brought up from a certain location, many people believe that the wreck will be in this exact spot. Later, it turns out to have been some part of the wreck that fell to the ocean floor, during the wreckages passage to Cape Cod after the sinking. This is how John Fish, Herbert McElroy and Peter Sachs and Arne Carr from the American Underwater Search and survey were finally able to pinpoint the wreck site, along with the help of Dr. Richard Limeburner from the Woods Hole Oceanographic Institute. Their search for the *Portland* wreck had lasted over the twenty years searching completely on their own in their spare time, which there was not much of, due to the fact that they were always working.

Captain Blanchard had been promoted to command of the *Portland* only two weeks prior to its loss. Had a position been available years before, he had all the skills of a first pilot to have made him qualify for Captain much earlier. Only when a senior captain of the line either died or retired, were lower ranked officers promoted to the exalted rank of Captain.

The old adage still holds true; everything that goes wrong is your fault, no matter who did it. In looking extensively at the side scan images of the wreck, particularly at the joint where the long rod, which connected the rear of the walking beam, on top of the engine to the paddle shaft, they had to see if the shaft joint was broken or intact

On another view of the *Portland's* engine, the main walking beam on top of the engine seemed to be angled forward to the extreme. It looked like evidence that the single double acting piston in the engine was stuck on the bottom dead center, thus stopping the engine. By studying these images of the engine and photos taken of the ship in the 1890's, it matched exactly the position of the famous photo of the *Portland*, steaming outward down the main channel in Boston Harbor. The superb photo was taken by the noted photographer, Nathaniel Stebbins of the ship with all the flags flying on her return maiden voyage to Portland Maine.

This photo was used in his 1890's book, Coast Pilot. The fact that the two thousand horsepower engine lasted for twelve hours, pushing a two hundred and eighty-foot steamboat into the seas that were forty to fifty feet high, with one hundred miles per hour gusts under extreme strain, is a testament to the skill of the Portland Engine Company and its' workers. The reason for the extreme forward tilt of the walking beam was the long up and down travel of the piston in the cylinder. This was necessary to obtain the required horsepower to run such a large steamer with only a one-cylinder engine.

So many things happened that night that even Lt. Nicholas Halpine, a retired US Naval officer could not figure it out. He had the most extensive knowledge of Massachusetts Bay. Captain Blanchard, of the Portland, had been dammed in the press by people who mostly did not know him. The many sea captains, some who were close friends, supported Captain Blanchard. Captain Leighton was supposed to take the voyage; however, he verbally disagreed with Captain Blanchard as the night did not look good to him. He chose to take the train instead. All the major players of the line had only been in their positions a short time and were still getting familiar with their new duties.

The line had recently lost two of their most experienced Captains. They were Captain William Snowman and Captain Charles Deering. It was Captain Deering's funeral in East Boston that most of the company officials were heading to and away from their jobs. This happened at exactly the same time that the *Portland* sailed to her doom. After Captain Deering's funeral was over, company officials would be attending more than their share. The *Portland's* engine or machinery never gave any trouble in all of the time it was in service prior to the storm. During its annual overhauls, the ship would be floated onto a cradle, which was hauled out of the water. Everything was inspected and made right if needed. Only the best would do for safety reasons.

Just before the ship was returned to the water, prior to the storm, it was partially floated over a period of two to three days to allow the bottom planks of the ship to swell up and expand to make the ship watertight. The cradle would then be completely lowered into the water, thus the ship would be backed off the cradle and returned to service.

It would now have a new safety certificate from the United States Steamboat Inspection Service, good for another year. The question has arisen "why didn't Captain Blanchard run the ship into Gloucester, the nearest port". Doing so, would run the extreme risk of turning the ship around and exposing the overhanging paddle guards to forty foot waves.

Trying to enter Gloucester Harbor in the dark, with blinding snow and only fifty-foot visibility with Norman's Woe Reef on the left and Dogbar Shoal on the right side would have been suicidal. You could hear the Monday morning quarterbacks now if the *Portland* was wrecked on the shoals. "Why didn't they ride it out at sea"? Riding it out at sea, had been the company policy for years. Other steamship companies did exactly the same thing, as in the course of the steel-hulled *Horatio Hall*, which rode out and survived the storm of the century.

The story of the black cat moving her kittens ashore just before sailing, was told by workers on the pier. The question arises, how many drinks did the pier workers have before they came up with that one.

Over the years, this story has become embellished many times until it becomes a story of its' own. One of the passengers had become settled on board the *Portland* and was taking a nap when an apparition of her mother appeared to her warning her not to take the trip. She promptly awoke, gathering all her belongings and crossed the gangway just as the crew were about to lift it onto the pier. This story appears as factual due to the woman's prerogative to change her mind.

The following is part of a letter written by the Boston agent of the Portland Steamship Company to the general manager of the line, Mr. John Liscomb. *"Dear Sir, on the train I may have told you about this, but perhaps not, for I could not think of everything I told George Barton, the watchman, to watch for the ring of the telephones, sure about nine o'clock for you had sent word for Captain Blanchard to wait until nine at night for the weather report Captain Blanchard would not, for he was bound to go on time and that you would be wild to hear that he had not waited. George Barton will swear that I told him that and it only goes to show that I said to Captain Blanchard all a man could say to follow out your request for him to wait. I am trying hard to collect my thoughts regarding conversation of that night with Captain Blanchard. Little by little, these things come to me. I telephoned twice this morning to tell of the life preserver man who came to my house just as I was going to bed at eleven o'clock and now comes the Tremont with the papers to make news of. It is awful. I think that I can get around so as to come Monday, for I shall keep going night and day. The papers report some receipt found. I know that from the letter I got from Cromwell Promark and. I will get hold of all those things. I went through all the letters last night before going home. Yours Sincerely, Charles F. Williams, Boston Agent."*

In similar conditions, during the Portland Gale, even modern ships with all the latest navigational equipment would be stopped in their tracks by similar conditions. They would have to jog in place until the seas died down and they were able to proceed. The injured survivors from the *Portland,* who were pitched into the water would have started to feel the effects of the hypothermia; thus becoming frozen to death in as little as ten minutes.

Having been to the wreck site twice, I could not see land in any direction. I thought to myself, if anything happens this far out, there is absolutely no hope of rescue at all, for even if a ship saw the *Portland* and stood by them, by the time you were hauled aboard the rescue ship it would be too late, as you would be unconscious. Artificial respiration did not exist at the time.

Final Sinking Scenario

Captain Hollis Blanchard sailed on the best information that he had and his best decision. He never got credit for the fantastic feat of seamanship, just by keeping the ship afloat and pointed it into the seas for over twelve hours. That was a feat of seamanship in itself. During the investigation of all the wreck data, I came to the final conclusion that the superstructure became separated from the deck of the ship, causing it to lose steerageway and control, as it separated from the hull. The top deck of the superstructure broke off and floated just below the surface of the water. The main hull then spun into the waves, and turned slightly to starboard on the right side, took a huge hit on the port-side ripping off the paddle-box. It rolled to starboard an angle of thirty-five degrees. All the dishes in the galley fell out of the cabinets all over the right-side of the ship. It then settled back on an even keel, burying it in the downwash. as the hull of the ship disappeared underwater It did a downward death spiral on its way to the bottom five-hundred feet below. It settled on the bottom with it's hull still intact, with a slight list to starboard. The bow of the ship is pointing south towards cape cod. The whole sinking lasted only thirty seconds. The fact that the dishes are still on the deck today, proves that the ship never capsized, as had been thought for over one-hundred years If it had, there wouldn't be any dishes on the deck at all. In conclusion, this gives total closure to the lives of the many fine crew and passengers on the *Portland*. May they rest in serenity.

EPILOGUE

Dive Story
By Robert Foster, Dive Team Leader

Over the years, many theories were posed as to the final location of the *Portland* and several efforts were made to locate the wreck by means of dragging. However, the wreck was not located until 1989, almost 100 years after its sinking, in what is now Stellwagen Bank National Marine Sanctuary. An ROV survey conducted in 2002 by NOAA brought back the first images of the ship since it was lost. These images show the hull intact, along with most of the main decking, but all of the upper decks gone and a few large fishing nets caught in the paddle wheels and on the still erect walking beam. The exact location of the wreck was kept secret.

On August 13, 2008, Capt. Doug Currier reduced the throttles on the Donna III to glide over the calm waters above the final resting place of the *Portland*. My first thoughts were about how horrible the conditions must have been on that night nearly 110 years ago. The estimated height of waves at sea was over 30 and possibly as high as 40 feet (9-12 meters). For a beamy paddle-wheeler with a shallow draft, it was surely a violent nightmare for the passengers and crew before the ship finally succumbed. On this day, the conditions were perfect, a requirement for the dive we were going to attempt.

We all began going about the preparations for the dive which had become second nature to us after many "practice" dives on other wrecks in the area, in depths of 250-350 feet (76-107 meters). Diving the wrecks in Stellwagen National Marine Sanctuary required us to develop unique techniques to balance personal safety with the special rules of the sanctuary (no grapnels, no tie-ins, no lines left in place). We originally experimented with the use of a shot line on the nearby wrecks of the *Frank. A. Palmer* and *Louise B. Crary* at 360 feet (110 meters), but this did not allow us to use a safety line hung with additional decompression gas which we deemed necessary. We felt the safety line was preferable to putting safety divers in the water who would incur decompression obligations themselves. Therefore, our procedure involved dropping a heavy weight to the up-current side of the wreck and dragging it back until it rested against the side. An ingenious system of tuna balls and surface line designed by Capt. Currier allowed us to tie up the boat to our rig without putting vertical stress on it, and allowed him to drop the rig and maneuver if necessary with divers still on the line. At the end of each dive, the last diver disconnected the main line from the weight leaving a breakaway line attached that would part and leave only the weight behind upon our departure. This technique limited us to diving in near perfect conditions and cost us our first attempt at diving the wreck, as we were not able to make a positive connection. However we gained a complete set of bottom and decompression gases from the wreck to the surface. A safety diver remained on the boat with extra gas if needed, but typically assisted with photos or stage bottle retrieval.

On this day, our patience paid off and we placed the line successfully next to the wreck near the highest point on the depth finder. We had the advantage of detailed sonar scans to help us interpret the readings on the bottom finder, but from a "height" of 500 feet through currents of varying direction, we felt lucky to have made contact anywhere.

Slav Mich, Don Morse, and I dove an open circuit configuration, and stayed together as a team. Paul Blanchette and Dave Faye dove rebreathers and comprised the second team. The rebreathers were to splash last and were tasked with disconnecting the main downline from our weight at the end of the dive. On this first dive, all of us planned short bottom times (approximately 12 minutes including descent). Don, Slav, and I submerged together and finally headed down after months of anticipation. As we dropped deeper the light gradually faded, and below approximately 330 feet (100 meters) left us entirely. Slav had placed a strobe on the last safety bottle designed to be about 30 feet above the end of the line (or just above the wreck) and I anxiously watched for it as the line disappeared into the darkness. Our lights constantly scanned the water for the nets we knew would be tangled in the wreck. Finally, the strobe appeared out of the black, and almost simultaneously my light picked up something ahead and to the left. While I slowed my descent, I began to make out the twin engine stacks that I knew were still standing above the deck. As soon as I realized what they were I knew our position on the wreck near mid-ships, and swiveled my light to the right to see one end of the massive walking beam still resting in place where it had stopped nearly 110 years ago. I'd like to say that my thoughts drifted romantically at that moment to imagine the elegant ship and the tragic events of the storm, but about that time I felt as well as heard a loud concussion that told me one of my video lights had imploded, as had happened to me on a previous deep dive, and reminded me of just how deep we were. Everything then was about time, gas, and breathing rate. On subsequent dives to the *Portland* I was comfortable enough to take in the wonder of my surroundings, but on this first dive the significance of what I was seeing would have to wait to be processed during the long decompression.

We dropped down on the deck at 443 feet (135 meters) in an area that was littered with dinnerware. Plates, pitchers, mugs, and serving dishes lay scattered in all directions. My light illuminated the gleaming white ceramic, in sharp contrast to the dull color of the silt that covered the wreck. Directly below me, a stack of glass rectangles, approximately 2 x 3 feet and edged in copper, lay neatly untouched and unbroken. Small bottles of various hues sparkled where they emerged from under the silt. For a few minutes there was no reason to move as my eyes continually picked out new artifacts from the detritus caused by decay. Sea life consisted of the large colorful anemones that always cover these wrecks, large orange starfish clinging to both wreck and dinnerware, and tube worms poking up from the thick silt covering the deck. Bright orange rose fish darted in and out of the wreckage nearby, and several large cod could be seen inside holes in the deck. The cod moved slowly and regarded me curiously in contrast to those on wrecks visited by divers. I slowly swam away from the line and towards the base of the walking beam, following Slav's light and keeping pace. He was also in no apparent hurry to leave the immediate area as his light jumped from object to object. A jumble of unrecognizable wreckage is collected at the base of the huge wooden supports for the walking beam, and we drifted upwards towards the steel beam and our up-line. On the way I noticed a large net rising up in the water column that appeared to be originating in the area of the port paddlewheel floating above the deck and pushed by the current over the center of the wreck. Slave swam around the opposite side of the beam, and I began heading for the strobe on the line, where I could see Don's light as well. It was time to begin the long journey back to the surface. As we stopped for our first decompression obligation, it occurred to me that I couldn't detect any surface light at all at that point, even when looking straight up.

We completed two additional dives to the *Portland* over the next several weeks, the second just aft of the area covered in the first dive, where we obviously landed in the area formerly occupied by the galley. My video captured stacks of serving dishes and a copper-covered warming bench. A starfish appears to grasp a large mug sitting upside down on the deck. The third and last dive was on the stern of the wreck, and area now completely flat and featureless but remnants of the staterooms that once stood above, a scalloped sink, a marine toilet, a simple white soap dish. Off the stern in the sand and mud bottom lie hundreds of additional dishes, pitchers, and other artifacts to remind us of the people that once traveled on this grand ship.

The sanctuary protects the *Portland* and other wrecks within its borders, and no artifact collection is allowed. Therefore, this ship will remain untouched although will eventually succumb to inevitable decay in the sea. Its tragic human loss, mysterious final resting place and inaccessible location all contribute to the *Portland* being referred to by some in the press as the "*Titanic* of New England."

Sara's Story

As I stared at what my father had brought out of the trunk, my eyes, normally large, were now as big as saucers. A green metal box no bigger than 10" x 8" held years of a family mystery that, up until then, I had no idea existed. "Mumum was adopted by a Portland Maine doctor and his wife when she was very young. Her father, John Whitten, was working on the *Portland* at the time it sunk and all were lost. This is all we know about Mumum's past and her biological mother, father, and three brothers." This my father said to me back in 1978 while reading, tossing and burning years of his memories from an old musty steamer trunk stored in our cellar, that held pictures, WWII journals and unlimited amounts of letters. His past. Little did I know then what we would learn and how much time and people it would take over the next 31 years to help unravel some of this family mystery, and what questions would go forever unanswered. Ours is but one of many stories of lives that were forever changed by the sinking of the SS *Portland* in 1898.

It wasn't until we sold our Maine Inn on Chebeague Island in 1984 that Dad and I started going from town to town in rural Downeast Maine to see if we could find out what happened to Lettie Whitten, my great-grandmother, and my grandmother's 3 biological brothers, Edward, Percy, and one we didn't have a name for. My grandmother's name at adoption was changed from Letticia Audrey Whitten to Audrey Whitten Thompson. It was the Whitten family we were in search of. I had spoken so much about this that it sparked Dad's interest. I was hoping to find Dad some long lost cousins! I was the instigator of this quest, as Dad had too much respect for his mother Audrey, who had passed in 1963, and her adopted parents and family, the John and Mary Little Thompsons, to be searching into something she had never shared with him. But as I hadn't gotten my driver's license yet, Dad was more than happy to come along. Back then in rural Maine where the towns were so small, the vital records were kept in homes of those in charge of such statistics. We started up in Machias where we thought Audrey had been born. As we went from town to town and home to home, we found out we couldn't even see some of the ledgers, as adoption information was still to be kept from our eyes. But the clerks could let us know if they had anything and we did find a marriage record of Lettie and John Whitten in Millbridge. More towns and the State capital of Augusta did not bring us any avenues to pursue. We headed south along the coast checking records in each town looking for Lettie's death certificate up to the 1940s, as we thought Lettie could not have lived past then. But once again we were very wrong. Finally, when I got my driver's license and set off alone down east from where we had left off, I found her in Rockland Town Hall. Some hunch had made me check a more recent range of dates. Lettie had died November 1959 in Rockland, one month after her last grandbaby was born, me, in New Hampshire. She had lived so close to our family and no one knew about her. How sad I thought. I laid flowers on a grave that no one had probably visited in a very long time and wondered more about what her life had been like.

She had lived at the Home for Aged Women in Rockland, which was amazingly still in existence in 1984. On a telephone call to the administrator I was able to hear warm memories of Lettie relayed to the administrator by one of the oldest residents still there that had known her 25 years prior. This warmed by heart. All further research into finding the whereabouts of any living cousins for Dad now had to wait another 11 years. I was young enough to be misguided into thinking I had all the time in the world. We moved off the continental United States. I had no idea of what was going to be revealed to me in the next 25 years. I would learn of some of the hardships Lettie had endured after her husband John died from the *Portland* disaster. But to what fathoms of grief she endured only she would ever know, and I can only guess at. It was when I was living in Maine at the time of the *Portland's* 100 year anniversary of its sinking in 1998 that I learned about the oldest brother, John "Edward" Whitten. Newspaper articles of the anniversary event peaked my interest and I was interviewed by Mason Smith who was co-authoring the *Portland* ship disaster, "4 Short Blasts," soon to be published. A *Portland* television news channel also interviewed me for the event. I was at the studio sick with a cold that had taken my voice away, but by the time the evening news broadcast it, their sound engineer magically had given me my voice back. But, it was the article and interview with Portland Press Herald writer John Richardson that connected me with my grandmother's brother's family. Edward, who died in 1964, had married Alice Snow in 1938. Alice had 3 children from another marriage and it was one of these now grown kids, Margaret, that contacted me. Her daughter had read the newspaper article and remembered that her step-great-grandfather had been on the *Portland*. I spoke on the phone with Margaret and her brother Daniel and heard them talk about the visits they remembered as kids with Lettie in Rockland. Edward never had biological kids but they knew Edward had a brother who had lived in the Boston area. In *Portland* there were get-togethers commemorating the 100 year anniversary, as well as one at the Abyssinian Church on the East end of *Portland* where most of the *Portland's* crew had attended back in the 1890s. The church, long ago abandoned, was now headed from restoration and to become a landmark. Most of the congregation in the 1890s were the men my great-grandfather worked, lived, and braved the wild seas with. I felt a connection to that church and still do.

Ten years later, and the invention of social media brought Art Milmore into my life.. Walter Hickey amazed me by going to the records departments in Portland and Boston and unearthing more Whitten facts that pieced together some of the answers.

After the *Portland* disaster, in 1900, my grandmother's family were no longer living together. Lettie was living at the Portland Invalids House; my grandmother was a ward of her future adopted parents in Portland; and the 2 boys, Edward and Percy were living at the Goodwill Farm in Fairfield. Walter found there had in fact been 4 children living during the 1900 census, but by the 1910 census only 3 living. One brother, Percy, who had moved to Boston, died in 1975 with no record of having any kids. It would have been fun for Dad to have met his step-cousins from his Uncle Edward's marriage, and to hear their stories of his uncle and grandmother. But Dad would never know of them or Edward and Percy as he passed in 1996. Edward lived and died in Laconia, New Hampshire. Lettie in Rockland.

Percy in Boston. So close to where we lived. If the sinking of the *Portland* caused Lettie to end up in the Invalids House, the boys to be placed in the Goodwill Farm, and my grandmother to be adopted, I'm wondering how many other spouses and children of the *Portland* crew and passengers ended up in the same way; families torn apart as a result of the disaster. We'll never know what catastrophe caused the third boy to die between 1900 and 1910, but will be forever grateful to Dr. John and Mary Thompson for adopting my grandmother. I would have liked to have known something about John Whitten, but only know that he was working on a ship on the ocean and living in Maine. Perhaps the ocean is in our family, as his great, great grandson Joe, my son, also lives and works on the ocean.

Sara Fuller 2015

ACKNOWLEDGEMENTS

When one begins a quest lasting thirty-two years, it takes the help of many people across a number of different states.

First and foremost, Mr. Edward-Rowe-Snow. I began reading his books at age nine and met him at age fourteen. As we became friends, I became involved in helping him on his Portland Book until the information ran dry in the 1970's

Mr. & Mrs. Leonard Bicknell and family. Mrs. Bicknell was Edward-Rowe-Snow's daughter, Dorothy.

Mr. Leo Delorey, who taught me about building and fixing boats when I was in my early teens.

Mr. Gunter Blatt, a holocaust survivor, who taught me about all aspects of photography.

Dr. Craig Macdonald, Mr. Ben-Cowie-Haskell, Principal organizer of the Portland Expedition, Mr. Matthew Lawrence, MS Deborah Marx and Ms. Ann Smircina. These fine people are from the Jerry Studds Stellwagen Bank Marine Sanctuary.

The crew of the research vessel Connecticut from the University of Connecticut's National Underwater Research Center, run by Dr. Ivan Babb, and his associates.

Allen Gontz, Research Assistant from the University of Maine

ROV pilot Craig Bissell &, Nick Worobry, Operations Coordinator

Mr. Bruce Terrill, underwater Archeologist from NOAA Office of Marine Sanctuaries

Captains' Turner Cabiness & Dan Nelson of the RV Connecticut as well as Mr. Dennis Ambridge, electronic tech from the University of Connecticut

Mr. Peter Demarco, writer from the Boston Globe Newspaper for his excellent article on both the expedition as well as the wreath laying

Mr. David Clark of David Clark Productions as well as his crew; Jeff Streich, the cameraman and Ray Day, the sound recording for their great production of the television special "The Wreck of the *Portland*

The Historical Marine group; Mr. John Fish, Mr. Arne Carr, Mr. Herb McElroy and Mr. Peter Sachs from American Underwater Search & Survey for their fantastic Search and discovery of the wreck of the *Portland* off Gloucester MA

Quints' Florist of Quincy, MA for fashioning the wreath

Mr. Fred Corning of Hingham, MA for the banner "Steamer *Portland*" used during the laying of the wreath over the Portland wreck

Mr. William Mueller, an artist from Cataumet, MA for his awesome painting of the *Portland* leaving Boston, MA. Mr. Mueller was one of the technical assistants to me, as well as being the helmsman on the side-wheeler, *Alexander Hamilton*, on the Hudson River during the 1950's.

Mr. Conrad Milster, Steam Engineer at the Webb Institute in Brooklyn, New York was also an assistant engineer on the Steamboat Robert Fulton. He looked at the underwater video with me regarding *Portland's* engine.

Mr. David Truby from the Massachusetts Board of Underwater Archaeology, who took pictures of the wreath laying with my camera

Mr. Richard Limeburner of the famous Woods Hole Oceanographic Institute, who did the computerized current drift analysis that helped me understand how the superstructure drifted south by east as well as how the veins of wind concentrated the debris, instead of spreading it out.

Ms. Sue Scheible for her excellent Patriot Ledger article of February 2016 on the lecture and the storytelling of the *Portland*

The Boston Public Library, in Boston, MA for their archive sections help in finding and copying many newspaper articles on the *Portland's* loss and subsequent searches. These include Curator, Henry Schnell, Diane Parks, John Devine, Xiping Zhang, Dawn Riley and Nancy Walsh

Stacy & John Burm of Hingham, MA for their help in correcting my English.

Mr. Nathan Lipfert of the Maine Maritime Museum

Mr. Daniel Fenimore of the Peabody Museum in Salem, MA

Mr. William Quinn of Orleans, MA for help in photography as well as fine-tuning this book

Roger and Virginia Snowman for their help with the information of Captain William Snowman, the first captain of the *Portland*. Roger was the great-great-grandson of the Captain. Roger's daughter, Sally Snowman-Thompson, who became the first woman keeper of Boston Light also proofread my early manuscript to simplify it.

The Late Albert Emery of Quincy, MA for his gift of the Charles Williams album

The men of the boiler department of the Bath Iron Works, who found out for me where the nameplate was located on their old style 1800's boilers to identify the wreck.

Stephanie Philbrick and the staff of the Maine Historical Society in Portland, Maine for letting me copy over 500 pages of old newspapers of the *Portland* sinking Mr. Edwin Dunbaugh, Professor of Humanities, Hofstra University, and his staff for their extensive knowledge of steamboats, as well as their help in fine-tuning the technicalities of steam propulsion.

Steamship Historical Society

Dr. Howard Gottlieb & Mr. Charles Niles from the Boston University, Mugar Library, which houses the extensive Edward Rowe Snow collection in which I spent many hours poring over all of Mr. Snow's huge collection of material about the Portland blizzard.

Mr. Paul Corkum, of Abington, MA for his photo enhancements in some of the drawings of the *Portland* and its wreckage on the outermost part of Cape Cod.

Mr. Bruce Tarvers, chairman of the Truro Historical Society for showing me the actual large artifacts in their collection, as well as allowing me to photograph them.

The Provincetown Museum in Cape Cod Massachusetts for showing me some of their artifacts from the *Portland*.

Mrs. Elizabeth Murphy from the Tufts Library in Weymouth, MA for her help in the preparation of two articles regarding the *Portland* for the Weymouth News

Mr. Philip Smith, former president of the Weymouth Historical Society for his help in wearing the Society's lifejacket from the Portland so that I could photograph it in an entirely different way.

The editors of the Weymouth News for their excellent help in preparing the two *Portland* articles for their local newspaper.

The late Frederick Corey, Superintendent of the Mount Pleasant Cemetery in Rockland, MA for showing me the grave of Mrs. E. Augusta Lara's grave. Her body was recovered from the outer Cape Cod beaches, after the wreck of the *Portland*.

Ms. Denise Larson of the Sagadahoc Genealogical and historical research room of the Patten Free Library in Bath Maine for helping me to pour over microfiche records of the 1889-1890 records of the building of the *Portland* at the New England Shipbuilding Company.

Mr. Peter Blackford & Mr. Bill Wyers, Superintendent of the tugboat, Lou-Ann-Guidry of Bath, Maine, who took me upriver to photograph the entire site of the New England Shipbuilding Company.

Mr. John Artisan of the Penobscot Marine Museum in Searsport, Maine, steered me to the correct newspaper accounts containing the *Portland* information.

Captain & Mrs. Douglas Lee of the schooner, Heritage, in Rockland, Maine for the location of other schooner wrecks in Massachusetts Bay and their memorabilia of the Portland disaster.

Mrs. Michelle Cappaleni, a member of the Weymouth Historical Society, for proofreading the early drafts of this book.

Mr. John Forbes of Holbrook, MA, for his persistence in helping getting this book finished, when the project became overwhelming, as well as his many long hours of work on the Edward-Rowe- Snow Memorial, which sits in a prominent place at Fort Warren on Georges Island in Boston Harbor.

Ms. Britt Steen-Breeden bender, curator of the Centerville, MA Historical Society, for showing me the small powered wheel, which was salvaged from the *Portland's* wheelhouse.

Mr. David Moore, formerly of Weymouth, England, for his fine drawings of the *Portland* that were especially created for this book, as well as allowing me to photograph them at his drawing board.

Mr. Mason Smith and Mr. Peter Batchelder of Cape Elizabeth Maine for their help with letting me research segments of their excellent book, "Four Short Blasts", which was published in 1998.

The Hull Lifesaving Museum in Hull, MA for their excellent articles written by their Education Director, Mr. John Galluzzo regarding the lifesaving exploits of Captain Joshua James and crew.

The Scituate Historical Society, in Scituate, MA, for their help in understanding the magnitude of the Portland storm.

Mr. David Ball and Mr. Fred Fiestas for their fantastic book, "Warnings Ignored", regarding the devastating effect of the storm on the town of Scituate

The late Mr. David Woodward of the Cohasset Historical Society for letting me photograph their artifacts from the steamer *Portland*.

The Portland Harbor Museum in Portland, Maine, which hosted a two-day seminar in which I was invited to speak.

The Southern Maine Technical College for providing their auditorium for the above seminar.

Mr. Richard Cleverley, a noted historian, from Hull, MA, who allowed me to copy some of his photographs of Joshua James and his lifesaving crew.

Mr. Randolph Hole of the Centerville Historical Museum, also of Centerville, MA, for helping me photograph the "powered main steering wheel" of the *Portland*, as well as photographing their magnificent model of this ship, which are on permanent display at their museum.

Ms. Laura Russo of the Boston University Mugar Library, Special Collections Department for helping me obtain photos of the "twin manual steering wheels" salvaged from the *Portland's* pilothouse washed ashore in Orleans, MA (cape cod)

Ms. Joanne Gamble of "Historic New England" for her great help obtaining a close-up photograph of Captain Joshua James and crew from the Point Allerton Lifesaving Station, which later was named the Hull Station of the United States Coast Guard.

Mr. Paul Walker, the Tufts reference librarian, in Weymouth, MA for his help with giving me sources and with proofreading material.

Sophia Yalouris, from the Maine Historical Society for her help with ordering the cover photo which showed excellent detail in the painting of the Portland, done by the artist, Antonio Jacobsen, in 1891.

Marcie Bilinski, Don Morse, Dave Faye, Paul Blanchette, Slav Mlch, Bob Foster (expedition head), Ricky Simon, Phil Dubey, as well as Captain Doug

Currie of the dive boat Donna 111, the expert diving crew, who dived three times to the wreck of the Portland and for showing me their video footage of the wreck, which was awesome.

Victoria Stevens, from the Hull Lifesaving Museum for her excellent support, as well as her copy of the photo of Joshua James and crew in front of the Lifesaving Museum.

Thank you to the Boston Globe and the Boston Herald newspapers for their extensive coverage of the Portland disaster and its aftermath. This information is found at the Boston Public Library's microfiche section.

The editors of both the Patriot Ledger, The Mariner Newspapers

Mr. John Chatterton, of the television series, Underwater Detectives, for reading as well as making suggestions on my manuscript.

Mrs. Mickey Hamilton of Weymouth, MA, for her wonderful article in the Weymouth News on the story and book of the *Portland*.

Mr. Dennis Lehane for his proofreading and suggestions on my manuscript.

The late Alfred Schroeder of the Massachusetts Metropolitan District Commission for his many years of help with the *Portland* story.

Karen Johnson, of the Weymouth Senior Center, for her help in arranging for me to do my lecture on the *Portland* there.

Mr. Dennis Molony from the Rockland Civil War Reenactors group for his help with my *Portland* lecture as well.

My wife, Margie, whose expertise typing and editing, helped get this book done.

Peter and Kimberly Donadio for their extensive help on the book

Kevin Tighe of the Braintree Mass. Geek Squad for his great help at the end of the book

Walter and Judy Hickey for finding Sara Fuller and finding the rest of Sara's family.

Sara Fuller for her great help in writing the epilogue.

Arthur J. Milmore-Author-E Weymouth MA, JULY 2, 2004- August 3, 2011
Edited and typed by extremely patient wife of 39 years.

Type of Work: Text
Registration Number / Date: TXu001198117 / 2004-07-20
Title: And the sea shall have them all.
Description: 89 p.
Copyright Claimant: Arthur Milmore
Date of Creation: 2004
Rights and Permissions: Rights and permissions info. On original appl. in C.O.
Names: Milmore, Arthur

The Library of Congress
United States Copyright Office
101 Independence Ave., S.E.
Washington, D.C. 20559-6000
202-707-3000

BIBLIOGRAPHY

———

Four Short Blasts: The Gale of 1898 and the loss of the Steamer Portland, printed in 1998 by Authors Peter Dow Bachelder and Mason Phillip Smith-Provincial Press, Portland, Maine

Night Boat to New England by Professor Edwin Dunbaugh, 1992 by the Greenwood Press, New York

After The Storm, by John Rousmaniere, Chapter 5- The loss of the Portland, Mass Bay 1898 published by International Marine/McGraw Hill-Camden Maine-2002

Annual Report of the Operations of the United States Lifesaving Service, 1899, in the collection of the National Archives Office, Waltham, MA

John Perry Fish-Unfinished Voyages: A chronology of Shipwrecks in the Northeastern United States-Lower Cape Publishing Co. Orleans, MA-1989

Warnings Ignored-The Story of the Portland Gale, by Fred Freitas and David Ball, Scituate, MA-1995

Steamboat Lore of the Penobscot, an informal story of steam boating in Maine's Penobscot region, by John M. Richardson, Kennebec Journal Print Shop-1941

Yankee Magazine article, December 1989 "How they found the Portland" by Susan Seligson

Great Storms and Famous Shipwrecks of the New England Coast, by Edward Rowe Snow, Yankee publishing -1943

New England Sea Tragedies, by Edward-Rowe-Snow, published by Dodd Mead-New York-1960

Strange Tales from Nova Scotia to Cape Hatteras, by Edward-Rowe-Snow, Dodd Mead-New York 1949

The Vengeful Sea, by Edward-Rowe-Snow-Dodd Mead, New York 1956

True Tales of Terrible Shipwrecks, by Edward-Rowe-Snow - Dodd Mead Publishing New York- 1963

Fantastic Folklore and Fact, by Edward-Rowe-Snow-Dodd Mead- New York –1968 (found in Chapter 10-THE FIRST SEARCH FOR THE PORTLAND)

Steamboats and their Giant Engines, by Robert Whittier

Judge Nathan Webb-Opinion of Court and Final Decree in the matter of the libel or petition of the Portland Steamship Co, owner of the Steamship Portland for limitation of Liability-Portland Maine 1899- in the records of the National Archives Waltham, MA

The Perfect Storm by Sebastian Junger- Norton Publishing Co. New York 1998

The Station Log- Hull Lifesaving Museum Articles on the Portland Storm by John Galluzzo.

Expedition Whydah by Barry Clifford and Paul Perry Harper-Collins Publishing 1999

Shipwrecks around Boston by William P. Quinn –Parnassus Imprints, Hyannis, MA 1996

The Charles Williams Album, Boston Agent, Portland Steamship Company 1896-1898 (in my collection)

Steamboats and Their Giant Engines by Mr. Robert Whittier, Duxbury, MA

The Snowman Family archives by Mrs. Virginia Snowman-ongoing, Weymouth, MA

The Means Library, Hull, MA by Dennis Means- Great-Great Grandson to Lifesaver Joshua James and a good friend of mine

The Journal of the Stellwagen Bank Marine Sanctuary, Scituate, MA 2002 and 2003

The Personal Collection of William P. Quinn, Orleans MA

The Collection of the Cohasset Historical Society, Cohasset MA

The Collection of the Scituate Historical Society, Scituate, MA

The Collections of the Centerville Historical Society, Barnstable County -Cape Cod MA

The Peabody Museum of Salem-April 2000 Lecture by Dr. Robert Ballard who found the RMS Titanic

The Collections of the Penobscot Marine Museum, Searsport, Maine

The Collections of the Maine Maritime Museum, Bath, Maine

The Bath Daily Times Weekly Editions in the Collections of the Patten Free Library, Bath, Maine

The Maine Historical Society Collections, Portland Maine

The Collections of the Tufts Library, Weymouth, MA

The Collections of the Boston Public Library, Main Branch, Boston, MA

The Archives of the Massachusetts Humane Society, Boston, MA

The Collections of the Provincetown Museum, Provincetown, MA

The Collections of the Truro Historical Society, Truro, MA

Boston: Story of a Port, by W.H. Bunting, 1971

Joshua James Collection, Richard Cleverly, Hull Historical Society, Hull, MA

The Collections of the Hull Lifesaving Museum, Hull, MA

INDEX

Chapter 1:
    Gloucester Harbor
    N.E. Shipbuilding Co.
    Bath Maine
    Ansel Dyer
    Lewis Nelson
    WM. Pattee Designer
    Bath Daily Times
    Portland Engine Company
    Cape Cod
    New England
    Norfolk Virginia
    Bath Daily Times
    Steamer Portland
    Captain Thomas Snowman
    Captain William Snowman
    Captain Roger Snowman

Chapter 2:
    Portland Steam Packet Company
    U.S. Department of Commerce
    U.S. Steamboat Inspection Service
    Amos E. Haggert

City of Portland
Hurricane Deck
New York
Fall River Line
Mr. John Coyle
Mr. John Liscomb
Franklin Wharf
U.S. Government
Boston
Steamer Tremont
Steamer Inspection Service
India Wharf
Steamer Longfellow
Captain W.M. Snowman
Maine Steamport Design
City of Portland

Chapter 3:
    Lief Erickson
    Mayflower
    Captain Christopher Jones
    Cape Cod
    New England
    Canada
    Gulf of Mexico
    Labrador Current
    Bay of Labrador
    Sable Island
    Newfoundland
    Gulf Stream
    Great Lakes
    Atlantic Ocean
    Maine

Delaware
Portland Storm
Boston
Cape Cod
Atlantic Ocean
Portland
Captain W.M. Snowman
Penobscot
Captain Thomas and Sarah Snowman
Newburyport Massachusetts
Portland Press
Portland Steam Packet Company
New England
Captain John Liscomb
Steamer John Brooks
Steamer Tremont
Steamer Baystate

Chapter 4:
Edward Rowe Snow
John Poland
Camden Maine
SS Portland
Portland Steam Packet Company
Haymarket Square
India Wharf
West Roxbury Station
Horse named "Dan"
Mothers Friends Bertha, Etta, and Willard
Boston
Night Boats
Grand Trunk Railway
New York City

Franklin Wharf
Maine
New York City
Boston
Portland Steamship Line - formerly Portland Steam Packet Co.
U.S. Steamboat Inspection Service
Maine's John Poland
Baystate
Portland Line
Boston Harbor
India Wharf
Rowe's Wharf
Atlantic Avenue
Nantasket
Provincetown
Steamer Longfellow - Collision
Cape Cod

Chapter 5:
Portland Steamship Line
Steamer Bay State
Portland Line
Captain Deering
Captain W.M. Snoman
Bay State
Boston
First Pilot Hollis Blanchard
First Pilot Alexander Dennison
Portland Line
John Coyle Jr.
Boston Agent John Liscomb
Thanksgiving Day 1898
First Mate Edward Deering

Captain Charles Deering
East Boston
Captain Blanchard
First Pilot Lewis Strout
Purser's Mate J. F. Hunt
Portland Co.
India Wharf
National Weather Bureau
Great Lakes
Gulf of Mexico
North Carolina Coast
Middle Atlantic
Bermuda
New England
Agent Williams
General Manager Liscomb
Captain Blanchard
Captain Deering
Captain Alexander Dennison
Bay State
Charles Blanchard
Mrs. Carrie Courtney
Lewiston Maine
Edward Rowe Snow
Boston Harbor
Thatcher's Island
Rockport Massachusetts
Isle of Shoals
New Hampshire
Deer Island Light
Wesley Pingree
Captain Collins
Steamboat Bangor

President Roads
Captain W.M. Roix
Steam Mt. Desert
Graves Ledge
Gloucester
Steamer Portland
Captain William Thomas
Schooner Maud S.
Captain Blanchard
Thatcher's Island
Londoner Ledge
Captain A. A. Tarr -Keeper of Thatcher's Island Light
Mr. Snow
Captain Lynes Hathaway
Brockton
Captain Reuben Cameron
Schooner Grayling
Costin Flare
Captain Frank Stearns
Schooner Florence
Eastern Point
Gloucester Harbor
Captain D.J. Pellier
Schooner Edgar Randall
Captain Albert Bragg
Steel Steamer Horatio Hall
Portland Harbor
New York City
Boon Island
Thatcher's Island
Captain Bragg
Mass Bay
Cape Code

Provincetown
Wood Propeller Steamer Pentagoet
New York
Rockland Maine
Captain Orris Ingraham
Captain Frank Scripture
Captain Samuel Fisher
Race Point Life Saving Station
Peaked Hill Bars Station
Captain Benjamin Sparrow
Cape Cod District Superintendent

Chapter 6:
Captain Charles Martell
Boston Tugboat Channing
Nahant
Captain Blanchard
North Carolina
Cape Hatteras
N. E. Coast
N. E. Area
Fishing Trawler Andrea Gail
Sable Island
Newfoundland
Steamer Portland
Cunard Liner Lusitania 1910
Cunard Liner Queen Elizabeth 2
Dr. Robert Ballard
Murphy's Law
Steamer Katahdin
Portsmouth New Hampshire
Cape Cod
Captain Samuel O. Fisher

Race Point Life Saving Station
Captain Michael Hogan
Schooner Ruth M. Martin
Peaked Hill Bar
Steamer Pentagoet
Captain Hogan
Cape Cod
Pentagoet
Merchants and Miners Line
Propeller Steamship Gloucester
Captain Francis M. Howes
Atlantic Coast
Edward Rowe Snow
Cape Cod
Pollock Rip Shoals Buoy
Steamer Portland
Norfolk Virginia
Boston Harbor
Steamer Gloucester
Monomoy Island
Chatham
Vineyard Sound
Hyannis
Pollock Rip Shoals Buoy
Pollock Rip Lightship
Cross Rip Lightship
Nantucket
White Star Liner Olympic - Titanic's sister
Titanic
Nantucket Lightship
West Chop Light
Martha's Vineyard
Gay Head

Block Island Light
Steamer Portland
Captain Frank Leland
Schooner Albert Butler
Outer Cape Cod Beaches
Surfman B. H. Henderson
Peaked Hill Bars Life Saving Station
Benjamin Kelley
High Head Life Saving Station
Jamaica
Boston
Norfolk VA.
Schooner Albert Butler
Peaked Hill Bars Life Saving Station
Schooner Mertis Perry
Brant Rock Marshfield MA.
Captain Joshua Pile
Grand Banks
Mass. Bay
Brant Rock Life Saving Station

Chapter 7:
    Maine - New York
    Mass. humane society
    Lovells island
    George's island
    Cohasset
    U.s. lifesaving service
    Cape cod lifesaving stations
        Wood End
        Race pt.
        Peaked hill bars
        High head

Highland
Pamet river
Cahoons hollow
Nauset
Orleans
Old harbor
Chatham
Monomoy island

1901 monomoy pt.added
Buzzards bay,
hen and chickens lightship
Nantucket island
Handkerchief shoal lightship
Monomoy pt.
Hyannis
Pollock rip lightship

Delaware
Cape cod
Schooner *king philip*
Martha's vineyard
Nantucket
Cape cod
Schooner *henry r.tilton*
Hull mass.
Boston
Point allerton lifesaving station
Capt.joshua james
Point allerton
4 masted schooner *abel babcock*
Philadelphia
Boston
Boston light

Stony beach
Nantasket roads
Toddy rocks
Windmill pt.
Hull gut
consolidATED COAL CO.
POINT ALLERTON
Pt.allerton lifesavers
Hull peninsula
Three masted schooner *henry r.tilton*
Stony beach
Pt.allerton lifesavers
Barge #1
Barge #4
Schooner *calvin baker*
Tugboat *Ariel*
Boston lt.
Great brewster spit
Mass.bay
Provincetown harbor
Wilson line steamer *ohio*
Spectacle island
capt.wm abbot-pilot
Gloucester harbor
Boston's south shore
Schooner *frederick alton*
George's island
Schooner *lizzie dyas*
Maine

Chapter 8:
    Thatcher's Island
    2000 h.p. Engine
    Gloucester harbor

Cape cod'
Provincetown
U.s. lifesaving service
Cape cod beaches
W.h.bunting's book - boston story of a port
Boston bangor skipper
Steamer *horatio hall*
Capt.albert bragg
Maine steamship company
Portland
New york
Schooner *ruth m. martin*
Capt.michael hogan
Steamer *portland*
Cape cod
Capt samuel o.fisher
Race pt.lifesaving station
Steamer *pentagoet*

Chapter 9:
    Steamer *portland*
    Capt hollis blanchard
    Stellwagen bank
    Peaked hill bar
    Nauset
    Monomoy island
    High head station
    Steamer *pentagoe*t
    Schooner *addie snow*
    Steamer *portland's wreckage*
    Portland
    Portland steamship co.
    Boston

Gloucester
Cape cod
Boston
Portland
Provincetown harbor
General manager liscomb
U.s.revenue cutter service
Cutter dallas from boston
Cutter woodbury from portland
Portland steamship co.
Boston and portland
Peaked hill bar
Cape cod
Charles f.ward chatham correspondent for boston herald newspaper
Hyannis
Truro
Portland
Thomas harrison eames author of book steamboat lore of the penobscot
Hyannis
East sandwich
Sandwich
Buzzards bay
Boston
Boston herald
Boston agent charles williams
Portland steamship co.
Boston steamer *Longfellow*
Provincetown
Passenger e.h.cook
Peaked hill bar lifesaving station
High head station truro mass.
Boston herald commercial wharf tugboat
Provincetown

Agent charles williams portland steamship co
William peak of barnstable cape cod
Charles herson
Steamer *portland*
Provincetown
Orleans
Mayo's blacksmith shop
Dr.samuel o. Davis
Orleans
North grove st.mortuary boston
Orleans
Thomas harrison eames steamboat lore of the penobscot
Portland
Capt benjamin sparrow of the cape cod lifesaving station
Schooner *maud s.*
Stellwagen bank
Portland's lanterns
Thatcher's island
Capt.pellier
1924 rockland maine
Capt.charles carv
Fishing dragger harriet crie
Portland
Cape cod mass.
Eastern seaboard
Portland steamship co.

Chapter 10:
Portland steamship co.
Boston globe newspaper
Boston tugboats *gallison and chesterton*
Peaked hill bars reefs
Cape cod

And The Sea Shall Have Them All

Peaked hill bar
Steamer *portland*
Dr.joshua lewis
Mass.governor roger walcott
Cape cod
Nantucket island
Martha's vineyard island
Boston tugboat herald
Outer cape cod
Vineyard sound
Peaked hill bar
Fishing schooner maud s.
Mass.bay
Duxbury mass.
Maritime provinces of canada
Nova scotia
Newfoundland
Island of saint pierre
Cable ship *minea*
Steamer *portland*
E.r.snow
Cape cod
Mass.bay
High head truro
North africa
Rockland maine
Capt charles carver
Schooner *addie snow*
Diver al george
Salvage ship *Regavlas*
Lt. E.R.Snow
Cape cod highland lt.
Pilgrim monument provincetown

Race pt coast guard station
Steamer *portland*
Peaked hill bar buoy
Steamer *portland*
Mr.snow
Patriot ledger newspaper of quincy
Mass.bay chart
1967
Steamer *portland*
Historic maritime group of new england hmg
Bourne mass.
American underwater search and survey auss
John Fish
Arnie carr
Herbert Mcelroy
Peter Sachs
Diver Al George
Snow site
5 masted schooner-unknown name
Cape cod
Hmg
Woods hole oceanographic institute
Dr. richard limeburner
1898 u.s.weather service
Cape cod
Cape ann
Side-scan-sonar
November 1898
Wcvb television
Chronicle magazine
22 miles
Cape cod
John Fish

Engine #57
Bath iron works
Portland
John fish
Mr.snow
Mason smith
Cape elizabeth, maine
Peter batchelder
Book 4 short blasts

Chapter 11:
December
U.s. representative john fitzgerald
Secretary of u.s treasury
Steamer *portland*
Portland steamship co.
Nathan cohen
Attorney benjamin thompson
U.s.district court portland maine
Judge nathan webb
Portland engine co.
N.e.shipbuilding co.
U.s. steamboat inspection service captains merrit and pollster
Amos haggert superintendent n.e shipbuilding co.
George morse of the portland engine co.
Engine #57
Portland steamship co general manager john liscomb
Boston newspapers
TULE LIFE PRESERVERS
John liscomb
Boston
Tule belt style lifepreservers
Portland

Judge webb
Portland steamship co.delaware
1899.steel propellor ship governor dingly
Steamer *tremont*
Joy line
New york
Steel propellor steamer *calvin austin*
Steel propellor steamer *governor cobb*
1901 eastern steamship co.
Maine steamboat lines
C.w.morse
Steamer *bay state*
Captain Deering
Capt Craig
Capt. Blanchard
Boston.
Mr. Snow
Pilgrim monument museum provincetown mass.
Truro historical society museum
Peabody essex museum of salem mass.
Snow family
Cape cod
Edward-rowe-snow
Weymouth historical society museum at the tufts main library
Maine maritime museum bath maine
Steamer *portland*
New England
Portland
Mass.
Maine
Samoset resort rockland maine
Texas shipbuilding co.bath maine

Edward-rowe snow archives tb university mugar library special collections Smithsonian Institution Washington D.C.

New england shipbuilding co bath maine
Woolwich maine
Front st
Portland slipway
Steamer *portland*
Bath ironworks
Smithsonian photo
Kennebec river
Whydah
Capt.black sam bellamy
Barry clifford
Provincetown
Wellfleet
Mel fisher of florida
Atocha
Mass.state board of underwater archeology
Barry clifford's dive boat *vast explore*
Orleans
The *whydah* gally 1716
The *whydah* museum provincetown mass.
Commonwealth of mass.
Capt.cyprian southack
Cape cod mooncussers
U.S.lifesaving service national geographic society museum washington d.c
Pirate museum macmillens wharf provincetown mass.
Chatham
Monomoy island

Chapter 12:
    Mass.coast
    Scituate harbor
    South shore
    Boston
    Norfolk virginia
    Tugboat mars
    Barges daniel tenney, delaware.
    Minot's light
    Collamore ledge
    Cohasset-scituate border
    Fourth cliff lifesaving station
    Third and fourth cliffs scituate
    Marshfield
    Capt.fred stanley
    U.s.lifesaving service
    Surfman richard wherity
    Book warnings ignored by david ball and fred freitas
    Henderson brothers
    Norwell
    Albert tilden
    George webster
    George ford
    South river
    Clapp brothers richard william and everett
    Thanksgiving
    Gundalow
    Fourth cliff lifesaving station
    Gulf of maine
    South river
    Snake river
    Henderson's
    Marshfield

Marshfield hills
Norwell
North scituate
Sand hills
Edward-rowe-snow
Suffman wherity
Storms and shipwrecks of new england e.r.snow 1943
Pilot boat columbia
Capt william abbot
Steamship ohio
Boston lightship
Pilot boat #21
Surfman richard tobin
North scituate lifesaving station
Schooner columbia
Surfman John Curran
Otis barker
Surfman tobin
Town of scituate
Scituate maritime and mossing museum photo 'pilot boat columbia'
Portland gale
North river
Fourth cliff scituate
Marshfield
Humarock
South coast of mass.
Tugboat mars
Boston harbor
British schooner Narcissis
Capt Macintosh
Boston, liverpool, shelburne, lunenburg, nova scotia
Seal island
Highland lt.

Truro
Ocean liner philadelphia
Barges #1 and 4
Point allerton
Tugboat underwriter
Boston
Boston herald newspaper
Four masted schooner king philip
Fall river
Cape cod
Camden maine
Cape cod bay
Gloucester
George's bank
Schooner gloriana
Grand banks
Race pt.provincetown
M. H. Walker
Essex 1889
Owner Edwin McIntyre
Captain Frank Miller
Schooner *Norman Fisher*
Captain John T. Dennon
Sloop *Venus*
Plymouth MA.
Clark's Island
Captain George Lewis
Schooner *New England*
Captain Lewis
Schooner *Gloriana*
George's Bank
Schooner *Edith Wilson, Whalen, Mizpah*
Schooner *New England*

Schooner *Shenandoah*
Gloucester
Grand Banks
Pubnico Nova Scotia
Gloucester Schooner *Emma Dyer*
Steamer *Herman Winter, Metropolitian Line N.Y. to Boston*
Nauset Beach
Orlean Cape Cod
Neil McNeil
Peter Hovel
3 masted schooner *Edgar Hanson*
Portsmouth N. H.
Dread Ledge Nahant
Nahant Bay
Swampscott Life saving crew
Captain Horton
Mattapoisett Schooner *Hattie Butler*
Angelica of Buzzards Bay
Captain Peter Mulven
Tugboat *Concord* Philidelphia
Barges *Satin Ella Taunton, Pioneer, Star of the East*
Hog Island Lightship
Prudence Island
Schooner *Nellie Craig* of Newport News VA.
New Haven Connecticut
Tugboat *Carbonero*
Boston Towboat Co.
Oilfield Point
Long Island Tugboat Herald
Barges Escort, McCauley, Naversink of Boston
Captain Hersey
Antonio Jacobson,
Dr. Ivar Babb

Nurc U.Conn Aquanaut Program
Katherine Zubrowski
Maine Historical Socitiey
Gloucester
Side-scan-sonar
Remote Operated Vehicle
Dynamic Positioning System
University of North Carolina
Captain Turner Cabiness
Bruce Terrill - underwater archeologist

Chapter 13:
Dr. Craig MacDonald - University of Hawaii
Mass. Gerry Studs
Stellwagen Bank Marine Sanctuary
Steamer Portland
Arnold Carr, Herbert McElroy, Peter Sachs, John Fish of American underwater Search and Survey
Cape Cod
NOAA - National Oceanic Atomospheric Admin
National Underwater Research Center - U Conn
Portland Gail
Steamship *Portland*
Boston
Portland Me
Research Vessel CT
Research Vessel Ferrell
Ben-Cowie-Haskell
Mass Bay
ROV - Pilot Craig Bussell
Portland Wreck Site
Steamer Portland
RV CT

Chapter 14:
   1889 - John Fish, Arnold Carr
   Scituate
   Steamer Portland
   HMG - Historic Maritime Group
   Stellwagen Bank Sanctuary
   Bourne Ma
   Dr. Craig MacDonald
   2000 HP
   Portland Engine Co
   Mass Bay Fishermen's Assoc.
   WCVB Channel 5

Chapter 15:
   Boston Agent Charles Williams
   Captain Blanchard
   Portland Me.
   India Wharf
   Steamer Portland
   President Roads - Boston Harbor
   Thatcher's Island
   Captain Michael Francis Hogan - Schooner Ruth M. Martin
   Manager John Liscomb, Agent Williams, Captain Blanchard
   Captain Fisher - Race pt Life saving station
   Steamer Portland
   Captain Albert Bragg
   Steamer Horatio Hall
   Provincetown
   Steamer Pentagoet
   Capt. Hollis Blanchard
   Edward Rowe Snow
   Boston Marine Society
   Boston

Portland Maine
Gloucester Harbor
Peaked Hill Bars
Cape Ann
Diver Al George
Highland Light
Schooner Addie Snow - Rockland Me
Mass Bay
Cape Cod
Lt. Nicholas Halpine
Boston Globe
Governor Roger Walcott
Cape Cod
Fishing Vessel Maud S
Tulle Life jackets

Chapter 16:
Steamer Portland
Captain Craig
Maine
Steamer Pentagoet
Boston Harbor
Steamship Ohio
Spectacle Island - Boston Harbor
Cape Cod
Charles Lincoln Ayling of Centerville
Hyannis
Edward Rowe Snow "New England Storms and Shipwrecks" 1943
Centerville Historical Society - Cape Cod

Chapter 17:
Mass. Bay
Stellwagen Bank Sanctuary

Steamer Portland - New England's Titanic
Portland Steamship Co
Cape Cod
A-Frame
John Fish and Arnold Carr
Portland Engine Co.
ROV
South America
Maine

Chapter 18:
Portland's john whitten watchman of millbridge maine
Letetia whitten
Audrey whitten
Mrs.whitten
Dr john thompson of portland maine
Casco bay
Portland maine
Whitten family
Portland abyssinian church
Munjoy hill
First parish church of portland
Miss emily cobb
Provincetown
Hull mass.cemetery-strangers corner
Portland storm
Portland steam packet co.
Steamer portland
Steamer tremont
Steamer bay state
Capt alexander dennison
Steamer gov.dingly.calvin austin.gov.cobb
Ne. shipbuilding yard

Texas shipbuilding co.
South portlansteamer portland
Merton l.small of woodfords, maine
Cambridge mass.
Boston
Woolworths dept.store
Mr.and mrs.arthur hersom
Arlington st.boston
Hammond beef co.portland maine
Albert carter
Mrs horace pratt and daughter any
Wm. roach of portland
Mrs a.s.chickering of weymouth mass. And sister e.augusta wheeler ofaugusta maine
Mrs. Mary Lara
Boston
Cape cod
Mrs. c.e harris
Abyssinian church portland
John Murphy and Oren Hooper and Son
Sophie Homes of Portland
Mercer Me.
Mr. F. A. Ingram - Percer of the Portland
Manager John Liscomb
Robert Foden - of Cincinatti
Portland Me
John Allen Dillon
Steamer Portland
East Port Me
Portland Me
Ms. Adge Ingram of MM Bailey's Store Woodford Me
Mayor Randol of Portland Me
Pro. W. R. Chapman of Portland Me

Chapter 19:
    Gloucester
    Research Vessel Connecticut
    Steamer Portland Wreck Site
    Dr. Ivar Babb - national underwater research center U Conn.
    Charles Williams Portland Steamship Co
    Science Channel
    Discover Channel
    Steamer Portland
    ROV
    Steamer Portland
    Edward Rowe Snow
    Wreath Banner - Steamer Portland
    Science Channel
    South Maine Technical College S. Portland
    Gulf of Maine
    Mass Bay

Chapter 20:
    Steamer Portland
    Atlantic Ocean
    Steamer Horatio Hall
    Steamer Pentagoet
    Steamer Dorchester
    Captain Albert Bragg of ME Steamship Co.
    Portland ME.
    Cape Cod MA.
    Boon Island
    Captain Hardin
    Cape Cod MA.
    Captain Harding First Pilot
    Nauset Chatham
    Pollack Rip Lightship

Shovelful Shoals Lightship
Handkerchief Shoal Lightship
Cross Rip Ligthship
Falmouth Basin
Vineyard Sound
New York
Chatham
Nauset
Race Point
Steamer Cambridge of ME
Captain Pierce First Pilot
Steamer Katahdin
Bangor ME
Boston MA.
Cape Porpoise ME
Portsmouth NH.
Boston MA
Nut Island Quincy MA
Steamer Forest City
Steamer Governor Dingley
Steamer Calvin Austin
Steamer Governor Cobb
Steamer Bay State
Eastern Steamship Line
German U-Boats
Chesapeake Bay Lines Steamer President Warfield
Exodus
Jewish holocaust
Middle East
Iserael

Chapter 21:
    Capt. Hollis Blanchard
    Atlantic Ocean

Collins Liner Atlantic
Halifax Nova Scotia
New York
Portland Maine
Great Lakes
Gulf of Mexico
U.S. East Coast
New England
Horatio Hall
Cape Ann
Atlantic Ocean
U.S. Navy Destroyer
Portland Steamship Line
Mr. Jot Small
Cape Cod

Chapter 22:
Steamer Katahdin
Maine Coast
Steamer Cambridge
Steamer John Brooks
Portland Argus Newspaper
Boston and Bangor Line
Capt. johnson
Penobscot river
Rockland maine
Monhegan island
Little egg rock
Steamer new england
Pemaquid pt
Engineer hawthorne
Steamer john brooks
Boston and portland line
Capt. liscomb

Mass, bay
Capt blanchard
Portland argus newspaper
Capt.jason collins
Steamer kennebec
Boston harbor
Capt.blanchard
Coal schooner king philip
Portland harbor
Peaked hill bar
Schooner maud s.
Portland steamship co.
Lt.nicholas halpine
Stellwagen bank
Capt william thomas of harpswell maine
Deering maine news and enterprise
Eastern pt light gloucester
Barnstable Cape Cod

Chapter 23:
Steamer portland
Boston bay
Cape cod
Boston globe
Portland steam packet co.Capt william bacon of tonawanda pennsylvania
Edward-rowe snow
Great lakes
Gulf of mexico
Provincetown
National weather service taunton, mass.
Dr.harold edgerton-mass.institute of technology
Steamer portland
Stellwagen bank national marine sanctuary

Kline sonar labs of new hampshire
Remote operated vehicle
Hela rov
Rv connecticut
Dr.ivar babb
Gloucester
Mr.snow
Capt blanchard
Gloucester
Steamer portland
Portland gale
Stellwagen bank
Cape cod
Peaked hill bar
Truro cape cod
Portland line
John fish, herbert mcelroy, peter sachs, arnie carr of american underwater search and survey-auss
Dr.richard limeburner
Woods hole oceanographic institute
Portland wreck
Capt blanchard
Boston harbor coast pilot book
Portland engine co.
Lt nicholas halpine u.s naval officer
Capt blanchard
Capt leighton
Capt.wm snowman
Capt.charles deering of east boston
U.s.steamboat insp.serv.
Capt blanchard
Gloucester harbor
Norman's woe reef

Dogbar shoal
Steamer horatio hall
Portland steamship co.gen mgr john liscomb
    Charles williams boston agent
    George barton
    Steamer tremont
    Cromwell promark
    Portland gale
    Steamer portland
    Capt hollis blanchard

CHAPTER NOTES

Chapter 1 – Building the Maine Masterpiece
Bath Times, 1889; The Archives of the Patten Free Library, Bath Maine; The Archives of the Penobscot Maritime Seaport, Maine; Archives of the Maine Maritime Museum, Bath, Maine; Archives of the Maine Historical Society, Portland, Maine; Archives of the Snowman Family of Weymouth, Mass. on Captain William Snowman, First Captain of the Portland; Virginia Snowman of Weymouth, Mass. family historian.

Chapter 2 – Take Her to Sea
The Patten Free Library Archives, Bath, Maine; The Bath Daily Times; Archives of The Portland Engine Company, Portland, Maine who built the 2000 horsepower engine. The Boston Post – records of the U.S. Steamboat Inspection Service, National Archives, Waltham, Mass.

Chapter 3 – Cruising the Maine Coast
Records of the U.S. Weather Service, and the Blue Hills Observatory, Milton, Mass.; The Portland Press Herald.

Chapter 4 – A Childs Memories
Edward Rowe-Snow, Great Storms and Shipwrecks of New England, Yankee Press. 1943; Short Blasts by Mason Smith and John Batchelder, Providence Press, 1998. John Liscomb album in the possession of the

South Portland Museum, Portland, Maine; Edward Rowe-Snow Archives, Mugar Library Special Collections, Boston University; Edward Rowe-Snow Fantastick Folklore and Fact, Dodd Mead & Co., 1968. Steamboat Lore of the Penobscot, 1940. Author's experience on a U.S. Naval Destroyer in the Atlantic and Mediterranean Sea.

Chapter 5 – The Faithful Decision
Steamboat Lore of the Penobscot by David Moore, 1940, 3rd Engineer of the Queen Elizabeth II and a friend of the author and illustrator.

Chapter 6 – At Sea in Mortal Danger
Edward Rowe-Snow, Fantastick Folklore and Fact, Dodd Mead & Co., 1968. Sebastian Junger, The Perfect Storm, Norton Publishing House. Colin Simpson, Lusitania, 1962. Lecture by Dr. Robert Ballard on the Titanic, Peabody Library, Salem, Mass.

Chapter 7 – Joshua James Legendary Lifesaver
The Station Log, Hull Lifesaving Museum; Portland Storm Articles by Mr. John Galuzzo; Edward Rowe-Snow, Storms and Shipwrecks,1943; The Boston Globe and Boston Herald storm accounts in the Boston Public Library Microtext Department.

Chapter 8 – Struggling to Survive
Boston Story of a Port by W.H. Bunting Steamboats and Their Giant Engines, Robert Whittier, Self-Published. Archives of the Blue Hill Weather Observatory, Canton, Mass. 1898.

Chapter 9 – Dark Clouds Gather
Edward Rowe-Snow, Great Storms and Shipwrecks of New England, Yankee Press. 1943; Boston Herald; Boston Public Library Collection Steamboat Lore of the Penobscot 1940; Boston Globe article, Boston Public Library Collections.

Chapter 10 – The Storm of the Century
Arthur Liscomb, Album of his Grandfather John Liscomb, Chairman of the Portland Steam Packet Co., collections of The South Portland Museum; Edward Rowe-Snow, Fantastik Folklore and Fact, Dodd & Mead, 1968. Susan Seligson, How They Found the Portland, Yankee Magazine, 1990.; Four Short Blasts by Mason Smith and John Batchelder, Provincial Press, Portland, Maine, 1998.; Report of the Portland inquires, National Archives, Waltham, Mass.; the author's three trips to the New England shipbuilding site, eventually finding the remains of the exact slipway which it was launched from in 1889.

Chapter 11 – The *Portland* Inquiry
Four Short Blasts by Mason Smith and John Batchelder, Provincial Press, Portland, Maine, 1998.; U.S. Government inquiries of the Portland trials, National Archives, Waltham, Mass.; the author's three trips to the New England Shipbuilding site in Bath, Maine.

Chapter 12 – The Storm Still Rages
Warnings ignored by David Ball and Fred Freitas, self-published; Edward Rowe-Snow, Storms and Shipwrecks of New England, Yankee Publishing, 1943; The Boston Herald newspaper articles of November 30, 1898 and December 1898, the Boston Public Library microtext archives.

Chapter 13 – Filming the Sunken *Portland*
Chaos in Boston Harbor, Gerry Studds, Stellwagen Bank National Marine Sanctuary press releases by Dr. Craig McDonald, Mr. Ben Cowie-Haskell, and Dr. Ivar Babb of the National Underwater Research Center, University of Connecticut; Dr. Robert Ballard's lecture of the Titanic at the Peabody Essex Museum of Salem.

Chapter 14 – New Developments
Gerry Studds Stellwagen Bank Oceanographic Expeditions to the Portland wreck in 2003 and 2004; verbal information from a lecture by Mr. John Fish of American Underwater Search and Survey, Cataumet, Mass.

Chapter 15 – Myths and Realities

Warnings ignored by David Ball and Fred Freitas, privately published.; Edward Rowe-Snow, Storms and Shipwrecks of New England, Yankee Publishing, 1943 in the author's own collection; Edward Rowe-Snow, Fantastic Folklore and Fact, Dodd Mead, 1968. Four Short Blasts by Mason Smith and John Batchelder, Provincial Press, Portland, Maine, 1998; Charles Williams personal album, author's collection; Arthur Liscomb, grandson to John Liscomb album at the South Portland Museum, Portland, Maine.; Testimony of Captain Albert Bragg of the Steamship Horatio Hall, which was hove to off Provincetown at the later; U.S. Government inquiry into the loss of the Portland; National Archives, Waltham, Mass.; News accounts of several sea captains in many newspapers as to what they would have done given the same circumstances as Captain Hollis Blanchard of the Portland.

Chapter 16 - The *Portland* Victims - The Human Cost

Captain Craig's report New England Storms and Shripwrecks 1943 - Edward-Rowe-Snow; The lost to the African-American community in Portland, ME; Munjoy Hill church closing 1914. Authors' personal recollections from Mr. Snow.

Chapter 17 – The First Search for the Portland

Boston Herald, Boston Globe, and Post accounts of the disaster, which was extensively covered in all local and many national newspapers; Author's recollections from going on two oceanographic expeditions to the Portland wreck, and viewed ten hours of live ROV footage on the wreck 500 feet below; The sudden failure of the Portland Abyssinian Church, on Munjoy Hill in Portland where many of the African American crewmembers on the Portland and who perished in the disaster. They made up most of the directors of the Church when the Church failed in 1914. Gladly it is being rebuilt today for its historic significance. The Hull Times and the Station Log of the Hull Lifesaving Museum. Unfinished voyages by John Perry Fish of American Underwater Search and Survey of Cataumet, Mass. on Cape Cod.

Chapter 18 – The New Portland Searches, and Victims' Stories
Sara Fuller Story, The Portland Press Herald, 1998. Her story has now reached its final, successful conclusion.; Calls to Walter Hickey at the National Archives, Waltham, Mass. about Sara, and his most complete list of the people who lost their lives that fateful night on the Portland; Portland Press Herald and Portland Argur stories at the Maine Historical Society Library, Portland, Maine.

Chapter 19- Expedition to the Portland September 13-18, 2003
The Stellwagen Banknotes, October and November 2003; author's recollections of the expedition on which he took very detailed notes, which later led to the information of exactly why and how the ship sank - the evidence is right on the deck of the wreck; The Science Channel high definition video of the wreck and their coverage of the wreath laying can be seen on television today, on both the Science and Discovery Channels, with the show title "The Wreck of the Portland." All forensic information was later analyzed to find the exact cause of the sinking.

Chapter 20 – Sidewheels Versus Propellers
Portland Press Herald, December 3, 1898; Captain Albert Braggs testimony at the Portland inquiry; Judge Webb's decision at the Portland Inquiry absolving the Portland Steam Packet Co. of any blame for the disaster, calling it "an act of God," Steamship Portland file, Maine Historical Society, Portland, Maine.; Unfinished voyages by John Perry Fish, Lower Cape Publishing Co., Orleans, Mass. 1989. Steamboat Dorchester, Edward Rowe-Snow filed Boston University Mugar Library Special Collections.

Chapter 21 – Laying the Blame
Edwin Dunbaugh, Night Boat to New England, Greenwood Press, Westport, Connecticut, 1992.; Engineering drawings, SS Portland, Maine Historical Society, Portland, Maine.; Four Short Blasts by Mason Smith and John Batchelder, Provincial Press, Portland, Maine, 1898.; Records of the

U.S. Maritime Steam Inspection Service, inquiry to the loss of the Steamer Portland, National Archives, Waltham, Mass.

Chapter 22 – The Sea Takes No Prisoners
Walter A. Lord, A Night To Remember: Story of the Titanic, 1950.; Portland Argus December 11, 1898 article on the Steamer Cambridge's struggle in a similar storm; Captain William Thomas of the schooner Maude S. letter written to the Deering Maine News and Enterprise, April 1899, John Liscomb Album, South Portland Maritime Museum.

Chapter 23 – Solving the One-Hundred Year Old Mystery
Boston Globe, Post, and Herald accounts from reporters on Cape Cod in December 1898.; Records of the United States Weather Service, Blue Hills Observatory, Canton, Mass. 1898, in which the wind speed gauges failed when the winds gusted to over one-hundred miles per hour; Boston agent Charles Williams' writing to John Liscomb, General Manager of the Portland Steam Packet Co., November 29, 1898 trying to explain Captain Blanchard's reason for taking the Portland out at 7PM, John Liscomb's letter, South Portland Maritime Museum. Author's two voyages to the wreck site in 2004 and 2004 witnessing a total of ten hours of live video of the wreck. The video DVD was later taken to Mr. Conrad Milster, Steam Engineer at the Webb Institute in Brooklyn New York. He had served as an oiler working on the walking-beam engine on the Hendrick Hudson and Robert Fulton. We watched the video together and he clarified for the author a burning question he had about a possible engine failure. To him the engine looked perfect in every way, and he identified a broken part laying atop the walking beam as a brace to the crosshead frame that had been deposited there after being ripped off by fishing gear. We also looked at an arc shaped image off the starboard bow that looked like a paddle-box that had been ripped off the wreck by fishing nets. He was a great technical advisor for the book, to which the author is forever indebted to.

Final Crew List of the Steamer Portland, November 27, 1898

This crew list for the final voyage of the Portland was reconstructed primarily with information obtained from the wage papers and claims filed with the U.S. circuit court. Information about race was taken from the U.S. population census of 1900 and the Canadian provincial censuses of New Brunswick and Nova Scotia of 1901. When census records were not available, that column was left blank.

Name	Rating	Birthplace	Race	Monthly Wages
Allen, Henry George	Porter	Harrisonburg, VA	B	22
Barron, Matthew	Deckhand	Placentia, Newfoundland		30
Berry, Mrs. Marjorie A.	Stewardess	Yarmouth, ME	B	18
Blake, Rodney S.	Watchman	Brooklin, ME	W	35
Blanchard, Hollis H.	Master	Belfast, ME	W	125
Bruse, Denis	Deckhand	N. Roms Island, Newfoundland		30
Carter, Allen	Fireman	Kouchibouguae, New Brunswick	W	35
Cash, William H.	Saloon	Wilmington, NC	B	22

Collins, Peter L.	Deckhand	St. John, New Brunswick	W	14
Cropley, George H.	Deckhand		W	30
Crozier, John	Deckhand	Willow Grove, St. John, New Brunswick	W	30
Daley. John	Deckboy	Cork, Ireland	W	30
Dauphinee, Everett	Deckhand	Chester, Nova Scotia	W	30
Davidson, James	Deckhand	DeBert, Nova Scotia	W	30
Dillon, John A.	Oiler	Eastport, ME	W	35
Doughty, William	Fireman	Eastport, ME	W	35
Dunn, William	Saloon			22
Dyer, Ansel Lewis	Qtrmaster & 2nd Mate	Portland, ME		35
Foreman, Lee Steam	Tableman	South Hampton, VA		25
Gately, John K.	Fireman	Portland, ME	W	35

Gatling, Alexander	Saloon	Elizabeth City, VA	B	22
Graham, George H.	Cabin Man	Burlington, NJ	B	22
Graham, Maurice	Deckhand	Simonds, New Brunswick	W	21
Harris, Mrs. Carrie E.M.	Stewardess	St. Mary's Bay, Nova Scotia	B	20
Hartley, Richard	Deckhand	Northeast, Newfoundland		30
Hemenway, William A.	Cabin Man	Worcester, MA	W	17
Heuston, William A.	Deck Steward		B	32
Howard, Stephen	Cook	St. John, New Brunswick	B	50
Ingraham, Frederick A.	Purser		W	90
Johnson, Charles H.	Saloon			22
Johnson*, Arthur A.	Saloon Watchman	Moose River, Nova Scotia	B	22
Jones, John	Cook	Frederick, MD	B	35

Latimer, William E.	Head SaloonMan	St. Croix, West Indies		24
Leighton, Franklin	Electrician	Falmouth, ME	W	50
Mackey, John	First Mate			45
Matthews, Alonzo V.	Steward		B	70
McGillivray, George	Deckhand	St. John, New Brunswick	W	30
McNeil, James J.	Oiler	St. John, New Brunswick	W	35
Merrill, Thomas B.	Chief Engineer	Norwalk, CT	W	90
Merrill, Charles L.	2nd Asst. Engineer	Westbrook, ME	W	50
Merriman, Hugh	Fireman	Harpswell, ME	W	35
Minott, Michael	Saloon	Port Antonio, Jamaica		22
Moore, Horace C.	Clerk	Portland, ME	W	17.33
Mundrucu, Theodore M.C.	Saloon Watchman	Pernambuco, Brazil		17
Nelson, Lewis Martin	Pilot	Norway		60

Norton, George A.	Deckhand	Lubec, ME	W	30
O'Brien, Con	Deckhand			30
Oxley, Ernest	Pantryman	Port Antonio, Jamaica		27
Patterson, Frank A.	2nd Mate	Belfast, ME	W	35
Pennell, Thomas H.	Fireman	Portland, ME	W	35
Pinna, Roland J.	Cabin Man	Cape Verde Islands		22
Reed, Griffin S.	Forward Cabin Watch	Portland, ME	B	22
Robichau, Winthrop P.	Baggage Master	New Brunswick	W	35
Rollinson, Harry C.	Fireman	Eastport, ME	W	35
Sewall, Thomas	Watchman	Westport, ME	W	35
Sloan, Arthur	Deckhand	Willow Grove, St. John, New Brunswick	W	30
Smith, Fred	Deckhand	Deer Isle, ME		30
Smith, Samuel Henry	Hall Man	Lynchburg, VA	B	22

Stanley, James	Deckhand	Brooklin, ME	W	30
Thompson, William G.A.	Cabin Man			22
Walton, John Tuck	1st Asst. Engineer	Cape Elizabeth, ME	W	60
Whitten, John C.	Watchman		W	35
Williams, James	After Cabin Watch	Bermuda		17
Wills, Fred A.	Cook	St. Martin, West Indies		25

IN MEMORIAM

———

Edward-Rowe-Snow mentor

Mr. Paul Blanchette master diver to the Portland

Made in the USA
Middletown, DE
25 March 2017